S. C. Ana

Bolton Ins

24th March, 1982.

Merrow Technical Library

Textile Technology

WOOL IN DOUBLE JERSEY

Merrow Technical Library

General Editor: J. Gordon Cook, B.Sc., Ph.D., F.R.I.C.

Textile Technology

MTL/TT/1	*Handbook of Textile Fibres*, J. Gordon Cook, 4th edition
MTL/TT/2	*Handbook of Polyolefin Fibres*, J. Gordon Cook
MTL/TT/3	*Spinning in the '70s*, P. R. Lord (Ed.)
MTL/TT/4	*An Introduction to Warp Knitting*, D. G. B. Thomas
MTL/TT/5	*Fluorescent Whitening Agents*, A. K. Sarkar
MTL/TT/6	*Self-twist Yarn*, D. E. Henshaw
MTL/TT/8	*The Setting of Fibres and Fabrics*, J. W. S. Hearle and L. W. C. Miles (Eds)
MTL/TT/9	*Weaving: Control of Fabric Structure*, K. Greenwood
MTL/TT/10	*Cotton Fibre Structure*, M. L. Rollins
MTL/TT/11	*An Introduction to Weft Knitting*, J. A. Smirfitt
MTL/TT/12	*Weaving: Conversion of Yarn to Fabric*, P. R. Lord and M. H. Mohamed
MTL/TT/13	*Wool in Double Jersey*, T. D. Brown (Ed.)

WOOL IN DOUBLE JERSEY

Edited by

T. D. Brown

Manager, Fabric and Clothing Technology,
International Wool Secretariat, Technical Centre,
Ilkley, Yorkshire, England

MERROW

Merrow Publishing Co. Ltd.,
276 Hempstead Road
Watford Herts England

ISBN 0 900 54179 2

Printed in Great Britain at the St. Ann's Press, Park Road, Altrincham

CONTENTS

1

INTRODUCTION

Knitting, as a means of producing articles of apparel, is of prehistoric origin. The early knitted articles were fully fashioned garments, but with the development of knitting frames in the late sixteenth and early seventeenth centuries, the production of knitted piecegoods commenced, both warp and weft knit fabrics being produced. Most of the weft knitted fabrics were what would now be known as single jersey, but also included both links-links and ribbed structures.

The development of the V-bed frame and the rib structure led eventually to the development of the interlock structure which was gradually developed over a period at the end of the nineteenth century. In 1908, two Englishmen, Scott and Wilkinson, took out a patent in the U.S.A. for an Interlock machine, and a monopoly position was maintained until 1928, when the patent expired. During this period there was some development of the machine, chiefly the use of two or even three cam tracks to extend its patterning scope. Swing clearing cams were introduced and this made possible some tuck interlock structures, the best known being Single Piqué.

In 1928, when the patent expired, other machinery makers became involved in the development of the Interlock machine, and important advances were made, notably the use of verge tricks. This permitted the knitting of some courses on only one of the two sets of needles which, together with the choice of either rib or interlock gating, made possible the development of other structures such as

Milano Rib. This was found to be in many ways an unsatisfactory structure, and a further structure—the now well-known Punto-di-Roma structure—was evolved. Other well-known structures, the Double Piqués and Bourrelet, were also introduced.

Development of these new structures led to an extension in the use of these types of fabrics. The original interlock fabrics were intended for underwear, but the newer structures were developed and used for outerwear, the phrase "double knit" being coined for outerwear in the period 1937–38. During the war years, the phrase "Single Jersey" became widely accepted; in post-war years, the double structures were revived and they became known as "Double Jersey".

It is interesting to observe that double jersey Jacquard fabrics were developed by a separate route, namely from single jersey Jacquard fabrics, which had been produced for a long time. In this development, the dial needles were introduced and used in order to avoid long floats on the back of the fabric.

Double jersey did not appear in any quantity until the mid-1950s, when it made a considerable impact on the women's outerwear market. At this time the women's tailored suit, which had been continuously popular since the mid-1930s, was running out of favour, despite the temporary boost given by the New Look in the late 1940s. The new double jersey fabrics gave a softer line than woven fabrics and were sufficiently different in texture and comfort to be an instant success in women's suits.

These fabrics were predominantly plain structures, and were predominantly wool, although some spun synthetic staple yarns—notably acrylics—were also used. Initially, continuous filament yarns were not very successful in double jersey fabrics. However, as techniques of texturizing developed, so did their use in double jersey increase and, by the mid-1960s, the predominance of wool and staple fibres was being seriously threatened.

These developments in continuous filament technology were being paralleled by other significant developments. One was the rapid increase in the ownership of domestic washing machines and the use of self-service laundries. This made it possible for the housewife to wash larger quantities of garments more easily than had previously been possible. Thus was evolved the novel concept of washing rather than dry-cleaning outerwear, the care of which thus became both cheaper and more convenient. This growth of the concept of washability and, in particular, machine washability placed wool at a disadvantage relative to synthetic fibres.

Other external developments were the expansion in the use of the private car and of domestic central heating. These together, did much to reduce the need to use clothing as a means of keeping warm—for which wool is pre-eminently suitable—and also resulted in a reduction in the weights of fabric used. Lighter weight fabrics were difficult to obtain in wool due to the cost limitation on raw materials, spinning, and knitting. Also, synthetic fibres for double jersey were marketed in bright, gay shades, providing added impetus to the boom.

It is scarcely surprising that, under this combination of circumstances, the use of wool in double jersey did not keep pace with the dramatic increase in the use of synthetic fibres, notably textured polyester. However, the use of wool in double jersey did not dwindle, but maintained a slowly increasing market. Meantime, the technology of wool made rapid advances, notably in the achievement of washability and the development of new dyeing techniques, giving faster, brighter shades.

Around 1971, the double jersey market in textured polyester reached its peak and attempts were made, with various levels of success, to extend it into men's and children's wear. However, it remained predominantly a women's wear fabric and, as such, more liable to fashion trends. It soon appeared that the impact of textured polyester would not last for ever, and in the early 1970s there was a trend back towards the staple fibre fabrics, including wool. At the same time, it was clear that the consumer would not abandon the generally accepted standards of design and easy-care hitherto associated with the synthetic fibres.

The underlying motive of this book must not be disguised. It is to assist knitters in the successful use of wool in double jersey fabric production and ultimately to increase wool's usage. It seeks to achieve this by making generally available the known technology of knitting wool in double jersey to the large number of knitters who may never have handled wool previously, and to show the techniques available for dyeing, shrink-resist treating, and finishing wool double jersey to meet the requirements of design and performance.

The purpose has been interpreted liberally, so that all stages of production from wool to fabric are discussed. This is done so that the knitter will be aware of his dependence on other stages in the production chain and so can adopt a realistic attitude to spinner and finisher. Likewise, it will enable others connected with the knitting industry to recognise the knitter's priorities.

By far the greatest part of the book is concerned with the technology of knitting wool. Wool is not the easiest of materials to knit, particularly into very tight structures, but the fabric has many desirable properties of comfort and appearance that are attractive to the consumer. This book should enable the knitter to master the techniques of knitting wool—which is already done on a large scale—and thus to gain access to larger markets.

Although much of the information is derived from investigational work carried out by the International Wool Secretariat Process and Product Development Department, information has also been obtained from many other sources ranging from practising knitters to academic researchers.

This was judged to be the right time to make such a book available for two reasons. On the one hand, developing technology appeared to have reached the stage where viable processes were available to produce attractive double jersey fabrics in wool with all the necessary easy-care properties; on the other hand, there is a revival of interest in the use of staple yarns in double jersey, which wool is now in a position to exploit.

2

YARNS

Wool yarns are classified as woollen or worsted according to the system of spinning, which influences the degree of orientation of fibres along the length of the yarn. The greatest degree of parallel fibre orientation is obtained in the worsted system and it is this type of yarn which is principally used in the knitting of double jersey fabrics. Worsted yarns can themselves be subdivided into oil spun and dry spun yarns, and again into weaving yarns and hosiery yarns. Oil spun yarns are rarely used in the production of double jersey fabrics as they contain about 3 per cent of added oil. This would remain in the fabric after finishing as double jersey fabrics, unlike woven fabrics and single jersey fabrics, are very seldom scoured (except in pale shades when solvent scouring is sometimes used to remove oil stains). The main stages of worsted spinning are shown in Fig. 2.2.

Weaving yarns differ from hosiery yarns with regard to twist and often the length of the fibre used, shorter fibres and lower yarn twists being generally utilized in the production of hosiery yarns. Double jersey yarns are, however, usually produced with higher twists than normal hosiery yarns, more nearly resembling weaving yarns in this respect although the shorter types of wool are used.

Although folded yarns are sometimes used in heavier fabrics produced on coarse gauge machines, the majority of double jersey yarns are knitted as singles, the most popular counts being in the region of 28 tex (32's worsted count).

Counts finer than this are difficult to obtain and to knit at a low fault rate.

The use of wool and other staple yarns requires a degree of forward planning which may be greater than for continuous filament yarn. Most double jersey fabrics are produced from top-dyed yarns and, although some spinners keep some colours in the form of roving or yarn stock, others only dye up tops on receipt of orders so that delivery times of the order of six to eight weeks are normal. An alternative, practised by some spinners, is to keep stocks of coloured tops and blend to give the required colour. This practice results in shorter delivery times for many standard shades.

The knitting performance of a yarn is influenced by the raw materials used in its manufacture and the processing

Fig. 2.1. Comparison of (above) woollen and (below) worsted spun yarns.

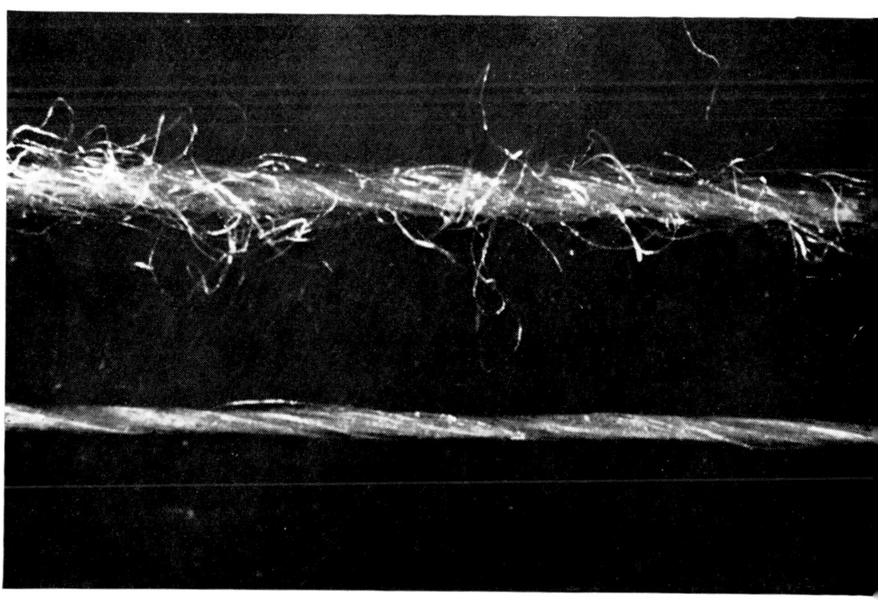

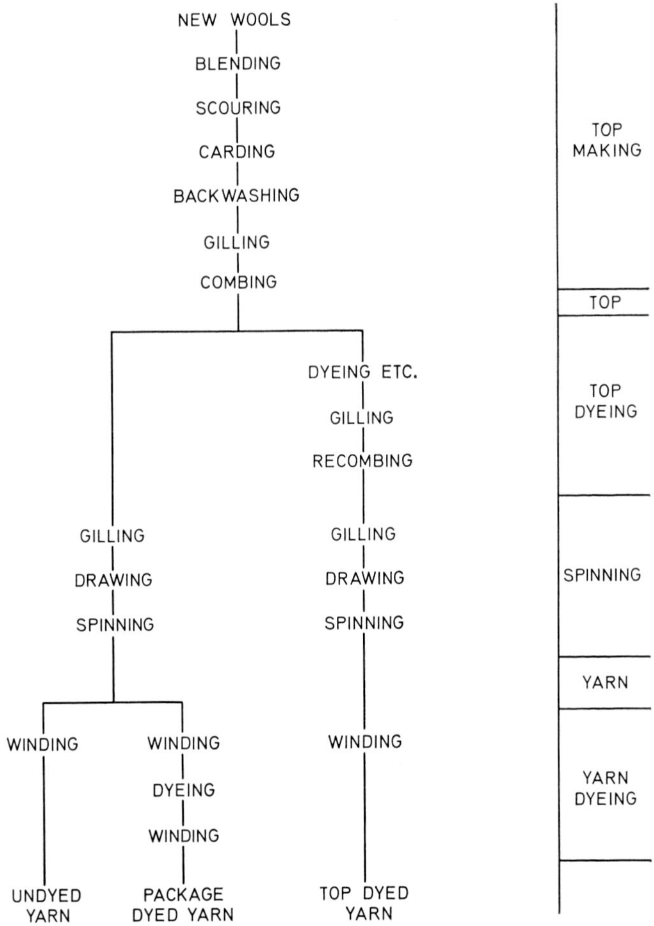

Fig. 2.2. The processing sequences for worsted spun yarns.

conditions utilized in the conversion of the raw materials to yarn. These are well developed technologies, and the knitter has much to gain from using reputable and experienced yarn suppliers.

2.1 Raw Materials and Top Making

The first stage in the production of worsted yarns is the manufacture of tops. There are many variables which affect the properties of wool fibres, such as the sheep's diet, age and breed, the climate, and even the part of the sheep's body from which the wool is taken. To overcome the difficulties raised by this sort of variability, it is usual to blend wools from different sources together. In this way it is possible to combine many different wool crops into blends which are very similar, not only from lot to lot, but also from season to season. The blend very largely determines the subsequent processing performance and even the properties of the final product. Thus, top-making is the crucial stage in worsted yarn production, requiring a great deal of expertise, and is often done by specialist firms.

Top-makers obtain their supplies of wool either from merchants or directly from the wool auctions. Whatever the source, wool is selected primarily on the basis of colour and fibre diameter, although other factors such as cleanliness are taken into account. Colour is of importance in the dyeing characteristics of the blend. While off-white or ecru wools are generally acceptable, wools which are yellowed would be unacceptable in blends to be dyed to light shades, but can be used where dark shades are to be produced.

Fibre diameter affects processing performance, end product type and properties, and to a lesser extent this is also true of fibre length. These aspects will be discussed in §2.1.1 and §2.1.2.

One important feature in top-making is the removal from the wool of all the grease and dirt which has accumulated during the period of growth on the sheep's back. After blending, the cleaning is accomplished by scouring, carding and backwashing. Scouring is done on the loose wool and removes most of the grease and dirt. Carding loosens up the wool and, being virtually a single-fibre treatment, it also aids the mixing process.

The sliver, or continuous rope of wool, which emerges from the card is subjected to a gentle washing to remove the dirt loosened in carding, and dried. Slivers are then fed in groups of four to six to gill boxes which straighten, mix and parallelize the fibres. This is done two or three times in order to prepare the sliver for combing, the essential element of worsted spinning.

Combing can be done either dry, or with the fibres lubricated by the addition of oil. Hosiery yarns are always dry-combed. Combing, as its name suggests, is a process in which the wool fibres are subjected to the action of combs. The purpose of this is two-fold, to render the fibres highly parallel and also to remove short fibres from the sliver. The short fibres removed are known as "noil', which is an important material for the woollen industry. The preferential removal of fibres is an important feature of combing. Shorter fibres are more likely to be combed out, but enough are left to preserve a continuous fibre distribution. Likewise, the noil also contains a percentage of long fibres.

The ratio of top to noil varies according to the long/short fibre balance in the top and is known as the "tear".

After combing, the top is finally formed by two or three further gillings of the parallelised sliver.

Tops are an important element in the worsted industry. They are sold by top-makers to spinners who may not produce their own tops. Tops are also often subjected to various chemical treatments, as discussed in §2.2. Blends of wool with other materials are often made at the top stage. For these reasons, the top is regarded as a stock item, and a starting point for many different production routes.

2.1.1 *Fibre Diameter and Variation in Fibre Diameter*

Fibre diameters may be measured by means of a microscope, but this method is tedious and very time consuming. Although the technique can provide a great deal of information about variation of fibre diameter, it is necessary to measure a large number of fibres if a reliable estimate of the

RELATIONSHIP BETWEEN BRADFORD WOOL QUALITY AND MEAN FIBRE DIAMETER

Wool quality	*Mean fibre diameter (microns)*
70	20·7
64	22·2
62	23·7
60	25·2
58	27·5
56	29·5
54	31·2
52	33·0
50	34·5
48	36·5

Fig. 2.3

mean fibre diameter is to be obtained. At the present time, fibre diameter measurements are often made using an air-flow method, which gives an estimate of the mean fibre diameter of a sample, but provides little information on variation of fibre diameter within the sample.

Fibre diameter and its variation are of importance for two reasons. In the first place, it sets a limit on the count or fineness to which a yarn can be spun; secondly, it affects stiffness, influencing, for example, the handle and drape of the resultant fabric.

Wool blends are usually referred to by a quality number. Originally this number was the finest worsted count to which the material could be spun on a mule. Since the development of the ring spinning machine for worsted yarns, with its different drafting and spinning stresses, the older spinning limits no longer apply and present ring spinning limits are coarser (i.e. lower worsted count numbers). However, the old quality designations have been retained, but are now defined in terms of mean fibre diameter rather than spinning limit. Wool qualities and corresponding mean fibre diameters are shown in Fig. 2.3.

The spinning limit arises because it is not possible to spin a yarn which has less than about 25 fibres in any cross section. Attempts to spin yarns having too few fibres in the cross section results in the production of excessively irregular yarns. Conversely, yarns having more than 25 fibres in the cross section are easier to produce with acceptable regularity, and commercially the limit is often set as high as 50 fibres in the cross-section. Since variations in fibre diameter will also affect the number of fibres in a cross-section, they also affect the upper limit to the counts that can be spun.

As explained above, the variation of fibre diameter is not often measured, but in the wools used for hosiery yarns, the coefficient of variation of fibre diameter is in the range of 20–25 per cent, the finer wool having the lower and the coarser having the higher coefficients of variation. By far the greatest proportion of double jersey yarns are produced from 64's quality tops. These have a mean fibre diameter of between 22·0 and 22·8 microns. Counts finer than 22 tex (40's worsted) are not usually spun from this wool quality, so that the yarns have a minimum of 40 fibres in the cross-section, and yarns of good regularity can be produced.

The diameter and variation of diameter of fibres also affect the drape and handle of fabrics. It is well known to engineers that the stiffness of a rod varies as the fourth power of its diameter. The effect of this can be quite dramatic—a 10 per cent increase in diameter results in a 50 per cent increase in stiffness—and needs to be borne in mind in designing a fabric. Fibre stiffness is not, however, the only factor involved in fabric stiffness, structure also being important.

Fibre stiffness is also an important factor in knitting performance, as it will affect the distortion of the yarn around the knitting elements and the ease with which the loops can be formed. This has not been studied in detail, but it is likely that any variation in fibre—and hence yarn—stiffness will affect the uniformity of loop shape within the fabric and hence influence fabric appearance.

2.1.2 *Fibre Length and Variation in Length*

Compared with fibre diameter, fibre length is relatively unimportant to the knitter; it is of primary interest to the spinner, who needs to adjust his drafting system in conformity with the fibre length. Like diameter, fibre length is not uniform but varies from fibre to fibre. Fibre length and its variation are controlled by blending and also by combing. Generally, fibre length tends to decrease with increasing fineness, i.e. thinner fibres are also shorter, but there is a great degree of variation. 64's quality hosiery tops have mean fibre lengths in the range 50–65 mm, and are frequently sub-classified by fibre length; the coefficient of variation of fibre length is about 55 per cent. Thus, the variation of fibre length is much greater than that of fibre diameter, but this is of importance in knitting only as it affects the regularity of the yarn. Shorter or more variable fibre lengths are more difficult to spin into a regular yarn. In addition, shorter fibres may require that slightly more twist be inserted into the yarn in order to maintain strength. This is discussed in §2.3.3.

2.2 Top Treatments

All the wool components of the final product are assembled during top making. The top is therefore a convenient stage for the application of various chemical processes, such as anti-felt treatments and dyeing.

2.2.1 *Anti-felt Treatments*

One of the causes of shrinkage of wool fabrics in washing is felting, which results from relative movement of the fibres in the fabric. This is due to the scales on the surface of the fibre and may be controlled by modification of the fibre surface, either by treatment with oxidising agents such as chlorine or by the application of resin (Fig. 2.4). An

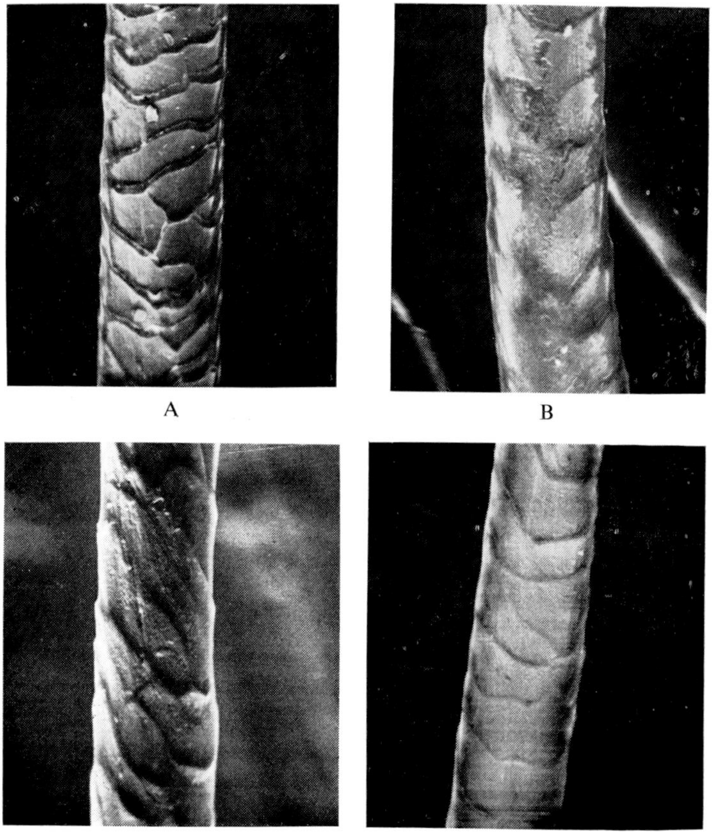

A B C D

Fig. 2.4. The effects of anti-felt treatments on the surface characteristics of the wool fibre. (A) Untreated fibre. (B) Treated by an oxidative anti-felt process. (C) Pretreated with chlorine prior to resin application. (D) Treated with resin after chlorination.

anti-felt treatment applied to the fibres will enable washable fabrics to be produced, but only if used in conjunction with treatments designed to control relaxation shrinkage. Alternatively, anti-felt treatments can be applied to the finished fabric. Methods of producing washable double jersey fabrics will be described in detail in Chapter 5.

An anti-felt treatment applied at the top stage can affect the knitting performance of the yarn. If an oxidative anti-felt treatment, e.g. Fichlor or Dylan X, is applied at an excessively high level, fibre damage can occur, particularly when the treatment is followed by dyeing.

This can result in a deterioration in knitting performance, particularly in fine count yarns. However, if the treatment is correctly carried out it is certainly possible to knit yarns up to 28 tex (32's worsted) count at an acceptable fault rate.

In the case of resin treatments (e.g. the I.W.S. "Superwash" process), in which the fibre is given a very mild chlorination in order to enable the resin to spread evenly on its surface, the knitting performance can actually be improved compared with untreated wool, although the actual knitting performance obtained is also dependent on the severity of the dyeing procedure.

2.2.2 *Top Dyeing*

Top dyeing is the most popular method of introducing colour into double jersey fabrics, as very fast dyes—which might dye unevenly on yarn or piece—can be used. Apart from Mélange or Vigoreaux printing (in which individual fibres are printed with bands of colour), top dyeing followed by mixing of coloured tops is the only way of producing mixture yarns.

Dyeing by any process before knitting results in a slight decrease in knitting performance compared with the knitting of ecru yarns, but the knitting performance of top-dyed yarns is generally considered to be acceptable. Any problems that occur are mostly in heavy shades such as blacks, browns, navies, etc., which have been subjected to the prolonged dyeing cycle used in the application of some chrome dyes, in which the dye is further bound to the fibre by the formation of a chromium complex. Over-dyeing is seldom done in top but, when it occurs, it is most prevalent in heavy shades and results in a deterioration in knitting performance.

2.2.3 *Recombing*

Felting and disturbance of the parallel orientation of the fibres inevitably occurs during top dyeing. Subsequent processing through pins (e.g. gilling) inevitably results in some fibre breakage. The resulting short fibres are not adequately controlled during drafting and can produce yarn irregularities in the form of slubs and thick places in the yarn which, in turn, can result in yarn breakage during knitting. Some of these defects may be removed in winding, but their removal necessitates replacement by a knot, and knots are also a cause of fabric defects, as will be discussed later (§2.4.3).

Short fibres can be removed in recombing, so that this usually has a beneficial effect on hole formation during knitting. Recombing is, however, a comparatively expensive process, costing about 4p/kg of top and is often omitted in order to reduce the cost of the yarn to the knitter. The cost of recombing should, however, be balanced against the cost of a hole in a knitted fabric and the general appearance of the fabric. Recombing is, of course, most desirable for the shades which have undergone the most severe dyeing conditions. Recombing is usual when coloured fibre mixes are being produced, where the final mixing must take place after dyeing.

2.3 Yarn Production

2.3.1 *Drawing and Spinning*

Starting from the top, which may have been subjected to one or more chemical treatments, including dyeing, the yarn is produced in five further stages. The first two stages are normally further mixing and parallelizing processes, known as doubling and drafting respectively, in which the weight per unit length is slightly decreased. The next two

Fig. 2.5. Drafting zone on a ring spinning frame.

stages are further drawing processes, often done without any further doublings, in which the weight per unit length of the sliver is considerably reduced. It is quite usual, although not essential, to introduce some twist into the sliver at the second, if not the first, of these two processes and the intermediate product is known as a roving.

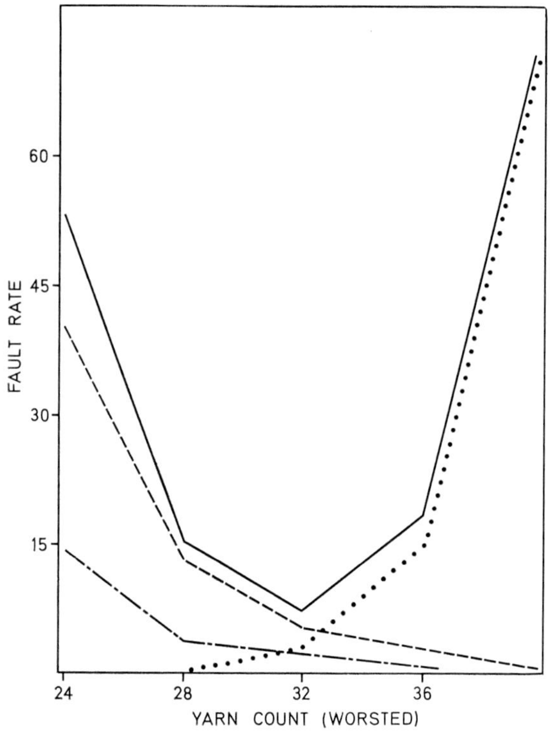

Fig. 2.6. The effect of yarn count on knitting rates.

——— *Total faults*

– – – – *Faults caused by knots*

· · · · · *Faults caused by thin places*

–·–·–·–·– *Faults caused by thick places*

By reason of the versatility of modern ring frames, it is possible to produce a wide range of yarn counts from any particular roving. Some spinners hold stocks of roving which can be spun up into yarn with a minimum of delay, thus keeping delivery times as short as possible.

Knitting yarns are spun on ring frames to a wide range of counts. Mule spinning probably produced the best yarns, but the process is now largely obsolete in worsted yarn production, due to its high cost. Cap spinning is also used, but is not generally recommended for knitting yarns as the yarn produced is very hairy. Yarn hairiness is discussed in §2.5.4. In spinning, the roving is drawn out to the final required yarn thickness or count and sufficient twist inserted to cause the fibres to bind together so that, under tension, the yarn will break with minimum fibre slippage.

For hand knitting yarns, knitwear and some single jersey yarns, two or more yarns are brought together and twisted to form a plied yarn. However, double jersey fabrics are nearly always knitted from singles yarns, and plied or folded yarns will not be considered here.

2.3.2 *Yarn Count*

Yarns for double jersey fabrics are produced in the range 22–44 tex (40–20's worsted count), although about 80 per cent of production is in the range of 28–32 tex (32–28's worsted count). These counts are easy and cheap to produce, easy to knit, and produce fabrics of acceptable weights, thickness, and tightness. Finer yarns, if spun, are more expensive, more difficult to knit, and will often not give the required fabric tightness or thickness. Coarser yarns are cheaper but are also difficult to knit and will produce excessively thick, heavy fabrics unless the tightness is considerably reduced.

The influence of yarn count is shown in Fig. 2.6 where the number of holes found in fabric is shown plotted against yarn count. It can be seen that in the case of coarse yarns, knots and thick places form the predominant causes of faults. In the case of fine yarns, the main cause of faults is thin (or weak) places in the yarn. The interaction of these effects results in the formation of a range of counts having a minimum total fault rate. This count range, corresponding to minimum fault rate, corresponds with the range in which the bulk of production occurs.

2.3.3 *Yarn Twist*

Yarn twist is the fundamental property affecting the rate of production and hence the cost of any yarn. Twist depends on, *inter alia*, the count and end usage of the yarn. In indirect or "count" systems the twist is proportional to the square root of the yarn count, the constant of proportionality, or "twist factor", depending on end usage and, as mentioned in §2.1.2, on fibre length. For the tex system, twist is inversely proportional to the square root of the yarn tex. Hosiery yarns are normally spun with the lowest possible twists, considerably lower than those used in weaving, and hence have the higher production rates. Yarns for double jersey knitting have, however, a much higher twist than other types of knitting yarns, twist factors in the region of 2,800–3,300 (tex) being common. Corresponding twist factors in other count systems are given in Fig. 2.7.

CONVERSION TABLE FOR YARN TWIST FACTORS

Indirect System

Twist = Twist Factor × Count$^{\frac{1}{2}}$

Direct System

Twist = Twist Factor/Count$^{\frac{1}{2}}$

Worsted system	*Cotton system*	*Metric system*	*Tex system*
1·8	2·21	67·8	2140
2·0	2·46	75·3	2380
2·2	2·71	82·9	2620
2·4	2·95	90·4	2860
2·6	3·20	97·9	3100
2·8	3·44	105·0	3330
3·0	3·69	113·0	3570
3·2	3·94	121·0	3810

Fig. 2.7.

Twist has an influence on yarn strength and extensibility and both these factors have some effect on knitting performance, as discussed in §2.5 below. High twist results in a firmer, more rod-like yarn and in the fabric this can result in a greater degree of stitch clarity, which may or may not be desirable, depending on the design and effect desired.

The fabric also tends to be a little thinner, more permeable to air, and perhaps also a little more abrasive to the touch, although this latter effect is minimized by steam finishing. Twist does not appear to affect fabric geometry, relaxation shrinkage, or stiffness.

2.4 Yarn Treatments

2.4.1 *Steaming*

It is usual to steam yarns in order to reduce the twist liveliness so that the yarn can be wound and knitted in a satistory manner. If it could be ensured that the yarn could be processed at very high moisture content, steaming would not be necessary and a slightly improved knitting performance could be obtained, as some decrease in knittability inevitably occurs during steaming. It is important, therefore, that the severity of steaming conditions should be kept to a minimum but, to do this, it is essential that the steam penetrates evenly throughout the package. To achieve complete steam penetration it is necessary to use a double cycle, i.e. vacuum, steam, vacuum, steam, vacuum. In the first steaming cycle it is only necessary to allow steam to enter the autoclave until the required steaming pressure is reached. When this is obtained, the second vacuum cycle may be commenced immediately. For the second steaming cycle, 2–5 minutes steaming at 70–100°C is adequate for most yarns.

2.4.2 *Package Dyeing*

Package dyeing of worsted yarns for double jersey knitting can result in an unacceptably high fault rate, particularly if the dyeing conditions are severe. The reasons for this high fault rate are not fully understood, but fibre damage is probably not the major cause. The major influences could be an increase in hairiness, loss of internal lubricants, and an excessive decrease in moisture content during the prolonged drying to which the yarn is subjected after package dyeing.

2.4.3 *Winding*

Winding is an important process as it gives an opportunity to remove yarn defects such as thick and thin places, and also is an ideal stage for the addition of knitting lubricants. Yarn count is the vital parameter influencing the clearing of yarn defects. For yarns coarser than an optimum count, knitting faults are mostly a result of thick places in the yarn, such as slubs and knots, and it is obviously inadvisable to replace a slub by a knot if the knot is thicker than the slub it replaces. Yarns finer than the optimum count mostly break at the thin places during knitting and, in this case, it is obviously preferable to detect and cut the thin places or to break them by winding at high tension and replace the thin places by knots, which cause very few defects in the knitting of fine count yarns.

Rubbing-up due to sharp edges on the guides during winding is an obvious cause of knitting defects. Another cause of defects and machine stoppages is "cob-webbing", caused by the yarn cross-winding over the base or nose of the cone.

Cones used for knitting yarns usually have an angle of 9° 15′ but, in order to obtain larger packages, there is some tendency towards the use of steeper angled cones. The combination of larger packages and steeper cone angles could cause greater tension variations, as the yarn is pulled off the cone. Excessive winding tension must be avoided, as this can lead to flattening and distortion of the yarn, adversely affecting fabric appearance. The yarn can also bind on the cone, leading to frequent machine stoppage during knitting.

2.4.4 *Lubrication*

As the frictional forces opposing yarn motion have a major influence on stress build-up, and hence yarn breakage during knitting, yarn lubrication is obviously an important feature in the preparation of the yarn for knitting. Paraffin

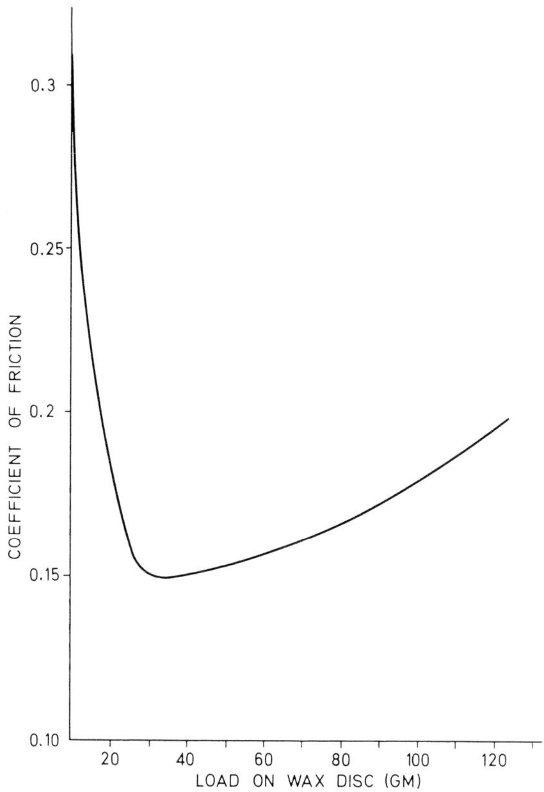

Fig. 2.8. The variation of coefficient of friction with waxing pressure.

wax is usually used for this purpose and is normally applied by wax discs during winding.

There are other yarn lubricants which are just as effective as paraffin wax, but none has been found which is better. Paraffin wax is therefore normally selected on the basis of economy. Paraffin wax is not a single material but a mixture and is available with a range of melting points. High melting point waxes are preferable as a yarn lubricant.

The amount of wax picked up is adjusted by the pressure of the discs on the yarn. As the disc-to-yarn pressure increases from zero, the resultant yarn coefficient of friction decreases very rapidly, and reaches a minimum value at a pressure corresponding to the weight of a $\frac{1}{2}$ in. (12 mm) thick disc. Thereafter, as the loading increases, the coefficient of friction also increases, but at a fairly slow rate. (Fig. 2.8.) Typically, an unwaxed yarn has a coefficient of friction of about 0·35 and this may be reduced, after waxing, to a minimum value of 0·15 (corresponding to the loading of a $\frac{1}{2}$ in. thick disc). For higher disc loadings, the rate of increase in the coefficient of friction is approximately linear at 0·05 per 100 g pressure. This relationship between friction and waxing pressure holds for yarns ranging in count between R110/2 tex (2/16's worsted) and 25 tex (1/36's worsted) and for a range of winding speeds and tensions.

It is thus apparent that excessive application of wax has an adverse effect on yarn friction. Also, the application of an excessive amount of wax can result in contamination of the knitting elements. For a typical knitwear yarn, the wax deposition for minimum yarn friction is about 0·3 to 0·5 μg (10^{-6} g) per cm, or about 0·05 per cent by weight. Double jersey yarns usually have an application of about 0·02 per cent of wax by weight. Theoretically, higher proportions should probably be necessary on double jersey yarns as these are much finer than normal yarns and so have a greater surface area per unit weight.

It is difficult to determine the wax content by solvent extraction of yarns, as wool yarns usually contain some oil which could lead to errors in the determination of added wax. In any case, it is the amount of wax on the surface of the yarn which has the major influence on yarn/yarn or yarn/metal friction, and measurement of the coefficient of friction is probably the best method of control. For a given wax addition, wax discs decrease the yarn/metal friction more than wax emulsions as, in the former case, most of the applied lubrication is on the surface. Wax emulsions have, however, the effect of increasing yarn strength, exten-

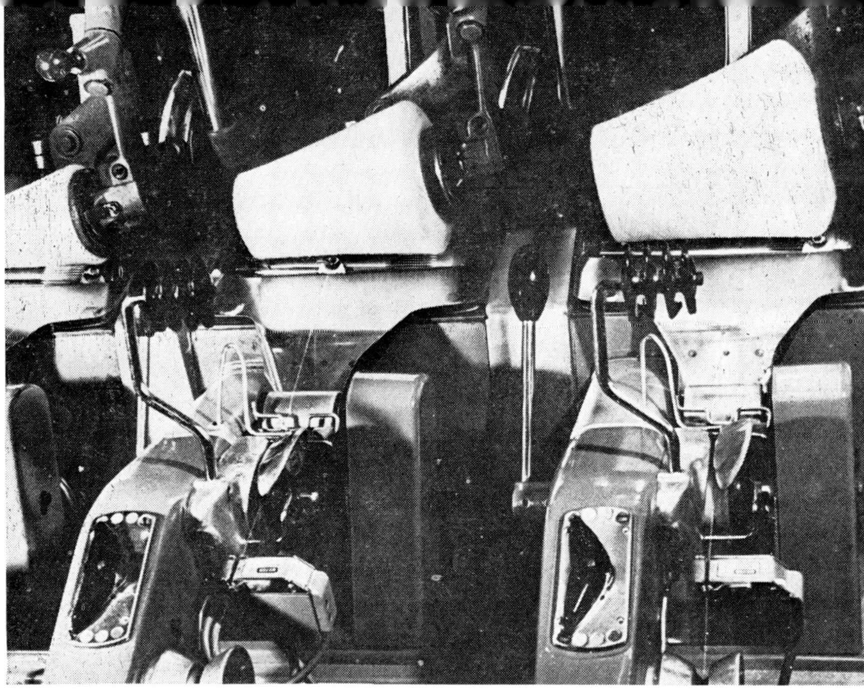

Fig. 2.9. The winding and waxing of worsted yarns.

sibility, and moisture content, so that the combined application of solid wax and wax emulsions is probably more effective than either method of application alone.

Although liquids are more difficult to apply than solids, developments in lubrication techniques and lubricants may result in the use of systems which give both effective internal and external lubrication of the yarn.

Variations in surface friction of yarns can also affect fabric tightness in those cases where positive feed is not used.

2.4.5 *Conditioning*

Wool yarns are normally sold on the basis of dry weight plus an additional allowance for moisture. This is expressed in terms of regain, which is the amount of moisture expressed as a percentage of the dry weight of the yarn. The standard allowance for moisture for double jersey yarns is $18\frac{1}{4}$ per cent, and although yarns may not contain this amount of moisture, an adjustment is made to the invoice in order to base charges on the nett weights at correct moisture content. Knitting yarns are usually delivered with between 12 per cent and 16 per cent moisture content, which is apparently sufficient for satisfactory knitting.

Moisture is usually applied by the spinner either by a centrifuge/spray system, by storing for long periods in a humid atmosphere, or by wet roller application during winding. This moisture assists in knitting as it increases yarn extensibility, and decreases the stiffness and the liveliness due to twist. It is important that the moisture be applied evenly throughout the yarn as variations in moisture content could result in streakiness in the fabric. A localized excessive moisture content can, in extreme cases, cause mildew to develop on the yarn. Indications are that wet roller conditioning, although inconvenient, can give the most uniform results.

2.4.6 *Storage*

In order to retain the moisture content of the yarn and to prevent damage to the outside of cones, the cones are usually wrapped individually with paper or polythene wrappings. Sometimes perforated wrappings are used in order to prevent an excessively high build-up of condensed moisture on the surface of the cone. The danger of using perforated wrapping is that the yarn could dry out if stored for long periods in a dry atmosphere, so that yarns should preferably be stored under conditions of high humidity. As the humidity for a fixed amount of moisture in the air increases with a decrease in temperature, it is preferable to store yarn in cellars, or certainly in a cool part of the knitting factory. This is only necessary if yarns are to be stored for prolonged periods, as a cone of yarn takes a long time to come to moisture equilibrium with the surrounding atmosphere. It is obviously important to avoid storage of unwrapped packages in the knitting room, as this often has a warm and dry atmosphere which will cause comparatively rapid loss of added moisture.

Humidification and temperature control in knitting plants for wool is probably desirable but, if sensible storage precautions are taken, no difficulties should be encountered due to drying out of yarns.

2.5 Yarn Properties

Of the principal yarn properties, count and twist have already been considered in §2.3.2 and §2.3.3. Rigidity was discussed in §2.2.1, and the effect of various top treatments in §2.2. The effects of various yarn treatments are discussed in §2.4, lubrication and friction being particularly important.

Strength, extensibility and irregularity also influence knitting performance and fabric properties. Other properties are thought to be of lesser importance.

2.5.1 *Yarn Strength*

Yarn strength is commonly supposed to be a major factor in knitting performance since it is the breakage of yarns which causes fabric faults and machine stoppages. However, it is now thought that the factor of importance is the occurrence of weak places in the yarn, and in this respect the statistics are a little intimidating. A roll of fabric weighing about 25 kg will contain about 800 km of yarn. Such a roll is likely to contain about ten holes. It is thus extremely unlikely that mean yarn strength alone will provide a means of predicting fabric faults. At the very least, knowledge of the variation of yarn strength is required which, together with mean strength, would enable an estimate to be made of the likelihood of a yarn break occurring. Thus, for example, a high strength yarn with a high strength variance is just as likely to have weak spots (i.e. produce fabric faults) as a weaker yarn with a lower strength variance.

It is often assumed, and it appears also to be true in

a general way, that the strength of a yarn is reflected in the strength of the resultant fabric. This is particularly true of fabric tearing strength since, in tearing, yarns are broken individually. Tensile and burst strength tests consist essentially of stressing large numbers of yarns simultaneously. However, the structure acts to redistribute the stresses between the yarns, and so structure plays an important part in determining the tensile and bursting strengths of fabrics.

2.5.2 *Yarn Extensibility*

Yarn extensibility is now thought to be a very important property in determining knitting performance. The argument put forward to support this is fairly complex, but is apparently sound. In the knitting zone of a double jersey machine, the first part of the knitting cycle is characterized by the relative separation of the dial and cylinder needles as they draw the new loops. The new loops are created by drawing yarn into the zone, either from the yarn package or by robbing back from the recently formed preceding loops. As the yarn runs into the zone, frictional forces are generated between the needles and the yarn. Thus the needles may be regarded as resembling the jaws of a yarn tensile strength tester stretching the yarn, but complicated by the fact that the yarn is also running through the test zone. If the yarn is insufficiently extensible to match the required increase in separation between the dial and cylinder needles, or the straining of the old loop as it passes over the closed latch at knock-over, then the yarn will break. (Alternatively, if the yarn is very strong, but with insufficient extensibility, the yarn will damage or break the needles. This does occur with, for example, continuous filament yarns.) Again, as in the case of yarn strength, both the mean value of extensibility and also the variation in extensibility are important. The reasons are very much the same; if values of extensibility below a certain value are to be avoided, this can be done by reducing the variation and also by increasing the mean value of the extensibility.

Yarn extensibility does not appear to play any great part in determining the extensibility of the fabric produced from the yarn. Generally knitted fabrics are much more extensible than their constituent yarns, and this is caused entirely by the deformation of the structure, which occurs without extending the yarn.

2.5.3 *Yarn Irregularity*

The part played by yarn irregularity in knitting performance, as in the case of other properties, has not yet been rigorously identified, but it is likely that its role is of some importance. Thick and thin places in the yarn have been shown in §2.3.2 to be particularly undesirable. Thin places in a yarn are frequently also weak places and would break on the knitting machine. They could be broken out in winding and be replaced by a knot, but this is not always satisfactory as knots are themselves an adverse factor in determining knitting performance and fabric quality. Knots and thick places can cause yarn breaks by interacting with the machine geometry. As a needle draws the new loop, the yarn is constrained to move in the restricted space between the needle and the wall of the trick. If the thick place or knot is too large, it will be unable to pass through this restricted space and will be held, jammed against the needle. This will result in the yarn becoming over-extended and breaking. This is discussed further in §4.1.

Thick and thin places, and indeed yarn irregularity in general, can have a marked effect on fabric appearance. Thick and thin places can result in irregular loop sizes and distorted loops. Yarn irregularity can give rise to stripiness, streaks or cloudiness, depending on the nature of the irregularity. Stripes would be caused by long term (between bobbin) variation, streaks by medium term, and cloudiness by short term variations in yarn count. Yarn irregularity is readily measured by the well-known Uster system, and standards for yarn irregularity are generally recognized.

Instruments also exist for classifying thick places in yarns by thickness and length. Opinions vary, but at the present time there does not appear to be any evidence relating such observations to either knitting performance or fabric appearance.

2.5.4 *Yarn Hairiness*

In general, yarn hairiness is a property which does not deviate greatly from accepted limits. Very occasionally, excessive variations in hairiness are found, leading to difficulties. In weaving, variations in yarn hairiness can give rise to a form of weft barriness, but in double jersey knitting this effect is very largely obscured by the multifeeder basis of yarn supply, which tends to obscure variation between yarns.

It has been suggested that in excessively hairy yarns, hairs from one section of yarn may be knitted into adjacent loops, bridging either wales or courses, although the evidence is slight. Hairiness does affect fabric appearance and possibly handle, hairy yarns producing a fuzzy surface appearance.

Yarn hairiness can have an important bearing on knitting performance. If the hairs are broken off the yarn in the vicinity of the knitting machine, they become airborne fly and will gather around the knitting elements and other machine parts. Accumulations of fly can and do become entrained in the yarn, resulting in blocked feeders and damaged needles, leading on the one hand to press-offs and on the other to needle marks. It must be stressed that such levels of hairiness are unusual in ring spun yarns, and are usually readily apparent to the experienced knitter when they do occur.

3

FABRIC STRUCTURE

Double jersey is a circular knit fabric produced by knitting on two sets of needles. Probably the simplest double jersey structure is the interlock, which is knitted on alternate cylinder and dial needles. Other plain double jersey fabrics, such as Milano Rib, Punto-di-Roma, and Piqué structures, consist of courses knitted on the cylinder or dial needles (plain courses) as well as courses knitted on cylinder *and* dial courses (rib courses). In Jacquard fabrics the yarns knit on alternate dial needles but knit on the cylinder needles only when required to show on the face. Blister fabrics are rarely knitted in 100 per cent wool yarns because of the excessive amount of strain imposed on the yarn during knitting.

3.1 Definitions

The quality of knitted fabrics is often described in terms of the courses and wales per unit length of fabric, but these vary considerably according to the state of relaxation of the fabric. The average length of yarn in a knitted loop does not, however, change appreciably with the condition of the fabric and so can be regarded as a fundamental parameter.

3.1.1 *Average Loop Length*

The *average loop length* is defined as the total length of yarn in the repeating unit or structural knit cell, divided by the number of loops contained in this repeating unit. The repeating units of some popular plain structures are shown by the dotted lines in Fig. A.2.1 of Appendix 2.

3.1.2 *Tightness*

The tightness of a double jersey fabric depends on the loop length and the yarn diameter, which in turn is dependent of the square root of the yarn count, i.e.

$$\text{Cover Factor} = \frac{1}{l\sqrt{N}}$$

where l is the average loop length in inches and N is the worsted count.

Tightness may also be expressed as "Tightness Factor" as follows:

$$\text{Tightness Factor} = \frac{\sqrt{T}}{l}$$

where l is the average loop length in centimetres and T is the tex count.

Tightness alone does not fully categorize a plain double jersey fabric, as the appearance, dimensions, and the physical properties also depend on the run-in ratio.

3.1.3 *Run-in Ratio*

The fabric run-in ratio is defined as the ratio of the length of yarn in a course when the machine is knitting on cylinder or dial needles (plain course) to the length of yarn in a course when the machine is knitting on both cylinder and dial needles (rib course). The term run-in ratio is only applicable to plain fabrics.

3.1.4 *Course Length*

Before a machine can be set up to produce a required fabric, it is necessary to translate the required cover factor and run-in ratio into course lengths. The length of a course is the length of yarn knitted during one complete revolution of the knitting machine. It depends, therefore, not only on the length of yarn in each loop, but also on the number of needles around the circumference of the machine. Course length is thus not a true fabric property, but is of great importance in the practical aspects of setting-up a machine to knit a particular fabric. Details of a method of calculating course lengths from fabric cover factor and run-in ratio are given in Appendix 2.

3.2 The Effect of Fabric Structure on Knitting Performance

There are many aspects of fabric structure which have an influence on knitting performance. Of these, fabric design is probably the most important. In the case of plain designs, the significance of both tightness and run-in ratio can also be recognized. In the cases of Jacquard and blister fabrics, the terms tightness and run-in ratio are not applicable, but obviously there are similarly variable conditions of equal importance.

3.2.1 *Design*

Knitting performance deteriorates with increasing complexity of fabric structure. Thus, a plain fabric can be knitted with fewer faults than Jacquard structures. At the other extreme of complexity, the knitting of double blister fabrics in wool is usually unsatisfactory from the point of view of knitting performance. The knitting of such fabrics is at least not recommended even if not positively discouraged. Single blister fabrics can be knitted in wool, but with a worse knitting performance than Jacquard designs. It is frequently necessary to reduce the machine speeds for blister fabrics, and occasionally for Jacquard fabrics also. The mending of fabric faults is an expensive process, and there is a shortage of skilled menders, so that it is frequently cheaper and easier to reduce the fault rate by reducing the machine speed than to repair the greater number of holes produced in patterned fabrics. This increase in fault rate and lower productivity, as well as design costs, explains the higher selling prices of patterned fabrics.

The increasing complexity of the structure in patterned fabrics can be related to retention of yarn on needles. In Interlock or Milano Rib, the yarn on any needle is replaced at alternate feeders, and in the other plain structures at every third feeder, as shown in Fig. A.2.1 of Appendix 2. Thus, Interlock and Milano Rib structures should be the easiest to knit. Interlock is a fairly popular structure, but Milano is little used, as the single rib course makes it highly vulnerable to a total loss of fabric from the machine in the event of a yarn break. The two Piqué structures and Punto-di-Roma are more popular structures since they are safer to knit than Milano and more stable than Interlock, although their knitting performance may not be quite as good.

At first sight, the Jacquard pattern structures appear to be very similar to the plain structures from the point of view of yarn retention on the needles. Thus in a two-colour Jacquard, loops are replaced on alternate feeders and in three-colour patterns on every third feeder. Only four-colour patterns, with loop replacement every fourth feeder, appear to be a more arduous structure to knit than a plain fabric. There are, however, two differences between plain and Jacquard structures which are of significance in this respect. Firstly, in Jacquard patterns, all the yarns are knitted into rib courses which are usually unbalanced since large numbers of consecutive cylinder needles may be knitted. This is potentially dangerous since it reduces the extent to which robbing back can take place and hence increases the stresses on the new loops as they are drawn. Secondly, there are short lengths of yarn, between alternate dial needles, which are not knitted on the cylinder needles, and these lie in the centre of the fabric. This bulk introduced into the structure will reduce the freedom of movement of the fabric in the knitting zone and hence will tend to increase the stresses on the yarns. This again will tend to increase the rate of formation of fabric faults.

Blister fabrics, like Jacquard fabrics, have yarn laid into

the centre of the structure. They are, however, further complicated by the presence of more than one loop on some of the needles. These loops are parts of different courses within the structure and thus represent a distortion of the structure. It is found that the loops can only be held in the needles by increasing the take-down tension, and this leads to a further deterioration in knitting performance.

3.2.2 *Tightness*

Fabric tightness, as expressed by a formal mathematical relationship for plain fabrics, depends on both the loop length and the square root of the yarn count. The effect of yarn count on knitting performance is discussed in §4.1 and will not be considered here. The other factor, loop length, is of considerable importance since it defines the amount of yarn carried on the needles, and hence the geometrical conditions in the knitting zone.

It is obvious, for example, that in a plain course the loop length cannot conceivably be less than the distance between needles, and there must be a similar minimum condition for rib courses. In practice, of course, loop lengths must be greater than these theoretical minima since the yarn must bend round the needles to form loops and be further restrained by the old loops and the dial and cylinder verges. From the static position, the loops have to be distorted by the needles so that they will pass over the latch at knockover.

It follows, then, that for any machine gauge and dial height setting, the shorter the loop length the less free length of yarn is available to accept distortion of the structure during the knitting action. This, then, throws greater stresses on to the yarn itself and thus increases the likelihood of yarn breakage.

The range of loop lengths, and hence fabric tightness factors which can be knitted, is also limited by the maximum and minimum loop lengths that can be knitted. The minimum loop length obtained in practice is that obtained when the needle depression is just sufficient to complete knockover and clear the old loop over the latch. Correspondingly, the maximum loop length is given by the loop drawn by a needle at maximum depression.

3.2.3 *Run-in Ratio*

In plain fabrics, the concept of a loop length has to be modified to that of an average loop length. Thus in plain structures the knitting performance is affected not only by the average loop length, but also by the actual length of the loops in the plain and rib courses. These lengths are governed by the run-in ratio so that extreme values of run-in ratio—either high or low—indicate small loop lengths in either the rib or plain courses with consequently increased knitting fault rates. Thus extreme values of run-in ratio will adversely affect knitting performance, and this effect is accentuated at high values of fabric tightness factor.

4

MACHINE VARIABLES

Double jersey knitting machines are available from a large number of manufacturers. Between them, the manufacturers offer a wide choice of pattern versatility, gauges, and number of yarn feeders. As a general rule, machines are offered with a certain patterning scope, together with some choice on both number of feeders and gauge. Machines are usually offered with a diameter of 30 in.

The simplest machines are capable of knitting only plain structures; more complex machines offer fixed selection Jacquard patterning ranging from 12 needle width (key selection) to 144 needle width (drum selection) pattern areas, based on mechanical selection. Electronic patterning systems are capable of producing non-repetitive patterns over the full fabric width. Whatever the degree of sophistication of patterning, only two distinctions apply in so far as knitting performance is concerned—either plain structures or Jacquard structures are knitted and the effect of structure has been described in §3.1. The methods of applying key, drum and electronic selection vary between machinery makers and so cannot be described here.

Double jersey machines are offered in a range of gauges up to 28 needles per in. Of these, 18 gauge is by far the most popular, with the installation of 22 gauge machines increasing rapidly. The proportion of other machine gauges in use is not great.

The vast majority of double jersey machines are of the rotating needle bed type. Overhead creels are general at

A

B

Fig. 4.1. Mechanical Jacquard pattern selection systems. (A) Pattern key system (fixed selection) used for small patterns. (B) A pattern drum system (variable selection) used for large patterns.

the present time, but free standing creels are being investigated in many places. Most machines are equipped with either a positive feed or constant tension storage device, these being fitted either by the manufacturer as an optional extra or by the customer after delivery of the machine.

The number of feeders on a machine has increased rapidly over the last few years, and 36 and 48 feeder machines are now commonplace. Increasing the number of feeders increases both the patterning area of the machine and the machine production rate, but its effect on knitting performance and fabric quality is indirect, depending on the modifications to machine design required to accommodate the increased number of feeders.

In setting up a machine to produce a fabric, it is necessary to bear two criteria in mind. In the first place, the knitting performance must reach as high a standard as possible. Failure to achieve this can result in an excessive number of machine stoppages, thus increasing production costs, and large numbers of fabric faults due to yarn breakages. In the second place, the fabric must meet its required level of tightness and have an acceptable level of stitch uniformity. Frequently, these requirements conflict and skill has to be exercised in order to achieve a satisfactory solution. In such cases, detailed understanding of the effect of machine settings on knitting performance is essential.

4.1 Machine Gauge

There is a very strong contemporary swing towards the use of finer gauge double jersey machines. The argument used in support of this trend is that use of a finer gauge enables tighter fabrics to be knitted making possible, in turn, the use of finer yarns to produce the lighter weight fabrics which are in demand. It is found that there is some relationship between machine gauge and the knitting performance of yarns. For example, in an 18 gauge machine the best knitting performance is obtained with yarns in the count range 28–32 tex (32–28's worsted count). Yarns coarser than

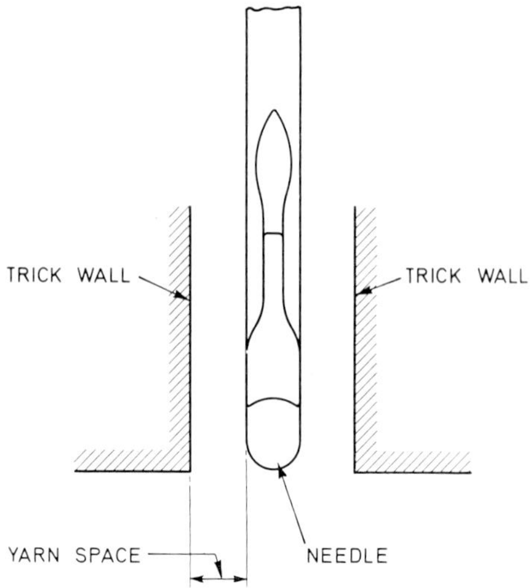

Fig. 4.2. The restricted space between the needle and the trick wall.

this produce faults due to thick places and finer yarns produce faults due to thin places, as described earlier in §2.5.3.

It is supposed by some authors that the optimum yarn count varies as the square of the machine gauge, but this is emphatically not the case. The supposition comes about by saying that the gap through which the yarn must pass, between needle and trick wall, varies inversely with machine gauge (see Fig. 4.2). This implies that the finer the machine gauge the narrower the gap. This gap, it is said, must be not less than the yarn diameter, which is inversely proportional to the square root of the yarn count. Thus machine gauge is proportional to the square root of the yarn count.

This argument suffers from one major fallacy which completely invalidates the conclusions. It is assumed that when the machine gauge is changed all the dimensions of the knitting elements, needles, tricks, etc., are changed in exactly the same ratio as the ratio of gauges. In practice this does not occur and, in fact, the gap between needle and trick wall varies very little between machines of different gauges. The errors involved in assuming this proportionality of machine dimensions are considerable. If it were true, the optimum count range for a 22 gauge machine would be 18–20 tex (49–44's worsted count). In fact it is nearer 24–28 tex (36–32's worsted count)—only slightly finer than for an 18 gauge machine.

The argument concerning the tightness of structure which can be knitted on a finer gauge machine is based on an argument similar to that used for optimum count. For a given yarn, the tightness of structure depends only on the length of a knitted loop and it is impossible that this be less than the length of yarn required to go round the needles on the machine. Thus tighter fabrics can be knitted on finer gauge machines only to the extent that the needles have been brought closer together. This decrease in needle distance applies to separation between dial and cylinder as well as between needles in the same bed. There is also a slight gain in the decrease in needle dimensions which may occur, but this is not usually very great.

Overall, the major effect of changes in machine gauge, particularly between 18 and 22 gauge, is in fabric width. Given a fixed machine diameter, the number of needles and hence the number of wales in the fabric width is proportionate to machine gauge. Provided that this increase in width can be properly exploited in the making-up, an increase in machine gauge can represent a substantial increase in the machine's production rate. For example, a change from 18 to 22 gauge represents a 22 per cent increase in fabric width, if the structure is kept constant.

4.2 Yarn Feeding

In the past, there was very little control of the yarn supply to the knitting machine. Efforts were made to control the

Fig. 4.3. Two yarn feed devices mounted on one machine. IRO constant tension units have been fitted above Triptape positive feed units. Courtesy IROPA (Textile Accessories) Ltd.

yarn tension by means of disc tensioners and carefully designed thread guides, and the yarn intake into the fabric was controlled by adjustment of the stitch cam setting. However, this control was not rigorous and variation in yarn input tension could cause fluctuations in yarn uptake, resulting in tight or slack courses which were readily observable in the fabric.

A number of devices are now available to assist in the supply of yarn to the knitting zone. The best known is probably the Trip Tape or Rosen positive feed system which draws the yarn from the supply package at a fixed but adjustable speed and forwards it to the knitting zone. Using this system, the rate of yarn uptake, and hence the loop length, is entirely controlled by the positive feed system and the role of the stitch cam is to adjust yarn tension.

Unfortunately, this system can only be used to provide two or three different yarn speeds so that the use of positive feed is, at present, restricted to plain fabrics.

For all fabric types, plain as well as Jacquard designs, it is possible to use constant tension units of which the IRO type is an example. These units draw the yarn off the supply package and rewind it in a standard form on an intermediate store from which it can be drawn off under standard, reproducible conditions. In this way it is ensured that the tension is held at any desired value and that fluctuations are kept to a minimum. Under these operating conditions the stitch cam is used to control the rate of yarn uptake, and hence the loop length in the fabric.

4.3 Yarn Input Tension

As indicated in §4.2, under present-day conditions it is possible to control yarn input tension to a fair degree of accuracy. This is a point of extreme importance as it controls the levels of yarn tension in the knitting zone. This is because the yarn tension at any point in the knitting zone, up to the point of knock-over, is given by the input tension multiplied by a factor for each needle over which the yarn has passed to reach that point. Since the factor concerned may be of the order of two for each needle, and five or six needles may be involved (see Fig. 4.4), the yarn tension may be up to 64 times the input tension. Under these circumstances it matters greatly whether the yarn input tension is 2 or 3 g as these values correspond to yarn tensions in the knitting zone of 128 and 192 g, the latter value approaching the breaking strength of many knitting yarns. In practice, it is recommended that, for wool, the yarn input tension should lie between 2 and 3 g.

It is not practicable to reduce the input tension to a very low level. Although this would obviously reduce the tensions in the knitting zone, there is also a tendency to lose control of the yarn over the first one or two needles in the knitting zone, thus leading to dropped stitches. In practice,

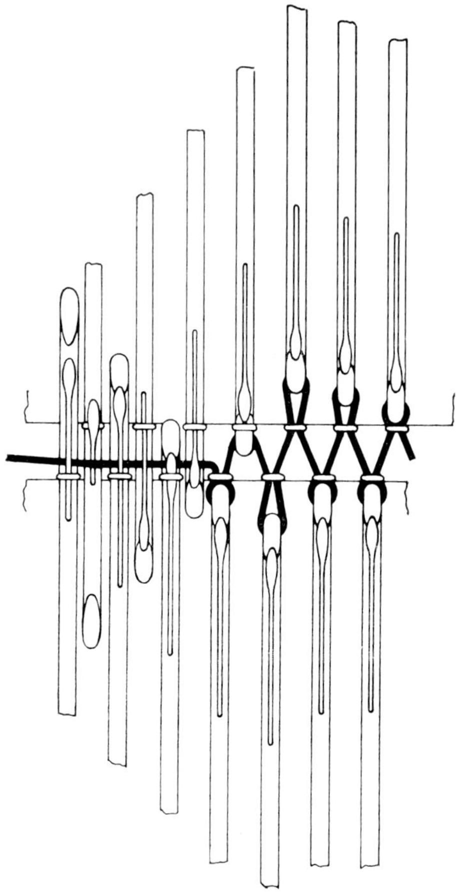

Fig. 4.4. The yarn path in the knitting zone of a double jersey machine.

a lower limit of 1·5 g is imposed on knitting tension.

A wide variety of yarn tension meters and speed meters is available which make the accurate setting of cams and positive feed units a simple matter for the machine operator.

As discussed above, the majority of creels for knitting machines are of the overhead variety. The use of side or free standing creels, if used in conjunction with positive feed or constant tension devices, need present no special difficulty in terms of yarn tension and input speed. It would be expected that attempts to use a free standing creel without the assistance of a driven yarn input system would be very hazardous due to the wide differences in path geometry, leading inevitably to differences in yarn tension.

4.4 Knitting Elements

It is a noticeable fact in knitting that many factors interact, and that some are of importance only in conjunction with others. This is the case with knitting elements, the relationship between needle and trick wall having been discussed in §2.5.3 and §4.1, where it was shown that this measure is of importance in determining both the tolerable level of thick places in the yarn and also the optimum count associated with a particular machine gauge. The other factor of importance is the dimensions of the hook, particularly over the closed latch.

As the new loop is being formed the needle hook, with latch closed, is forced through the old loop at knock-over. This results in considerable distortion of the old loop, and also of the loops on either side of it, particularly if the fabric is being knitted to a tight structure and the loops are small (Fig. 4.5). It will be appreciated that the old loop can be highly strained at knock-over and may break if the yarn extensibility is too low. Thus the size of the needle directly limits the tightness to which a fabric can be knitted, and fabric holes can be caused by trying to knit tighter than the size of needle permits. Since, in fact, needles are not normally scaled down in size on finer gauge machines, this restricts the change in fabric tightness obtained by changing to finer gauge machines.

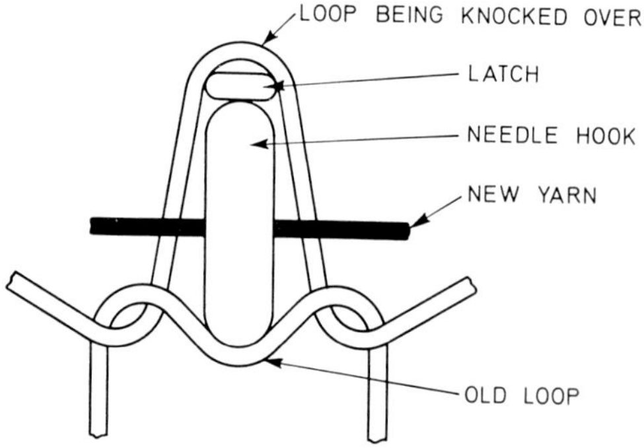

Fig. 4.5. Distortion of the old loop at knockover.

4.5 Cam Design

Much work of an academic kind has been done on the optimum shape of cams and some extremely interesting non-linear cams have been designed. Unfortunately these have all suffered from one or two disadvantages relative to the linear cams used on commercial machines. Firstly they are comparatively flat, extending over very many more needles than the linear cam, thus sharply restricting the number that can be fitted into a standard machine diameter and so reducing the machine production rate. It is, argued however, that non-linear cams would permit higher machine running speeds. Secondly they are not adjustable, so that changing the stitch cam setting could only be accomplished by replacing the cam completely. This would be grossly uneconomical both in terms of the changeover time required and in the investment required in spare sets of cams.

On machines of standard diameter, an increase in the number of feeders can only be accomplished by crowding the cams closer together around the machine circumference. This is done by making the cam angles much steeper.

of yarn available for the dial needles to draw is progressively decreased. This results in the knitting of a tighter fabric which, in some cases, may be highly desirable. However, the yarn is also more highly strained in the process of knitting this tighter fabric, so that knitting performance may be expected to deteriorate, resulting in a higher fabric fault rate.

This argument applies principally to plain fabrics where robbing back and other yarn motions can occur over adjacent needles. In Jacquard structures, delayed timing may be more difficult since yarn motion may not be over adjacent needles, depending on the design used, and yarn strains may become excessive. It does appear, however, that a delay of one or two needles is tolerable, even in Jacquard fabrics, thus increasing the tightness of these notoriously loose structures.

4.7 Dial Height

Consideration of the disposition of the fabric on the double jersey machine shows that most of the fabric weight and take-down tension is carried on the dial needles, which are at a greater height than the cylinder needles (see Fig. 4.7). Thus the loops on the dial needles are more greatly stressed than the cylinder loops, and this is confirmed by the much greater occurrence of yarn breaks on the dial than on the cylinder needles. Further, if the dial height is too great and positive feed is used (in a plain structure), then the yarn path between and around the dial and cylinder needles may be physically greater than the yarn fed into the knitting zone so that the yarn is under tension even in the "rest" position (i.e. after the new loops have been formed). For this reason, the dial height should not be used as a means of adjusting the yarn take-up tension. It is recommended strongly that the dial height should be as low as possible, consistent with allowing the knitted fabric to pass freely between dial and cylinder to the take-up below. Doing

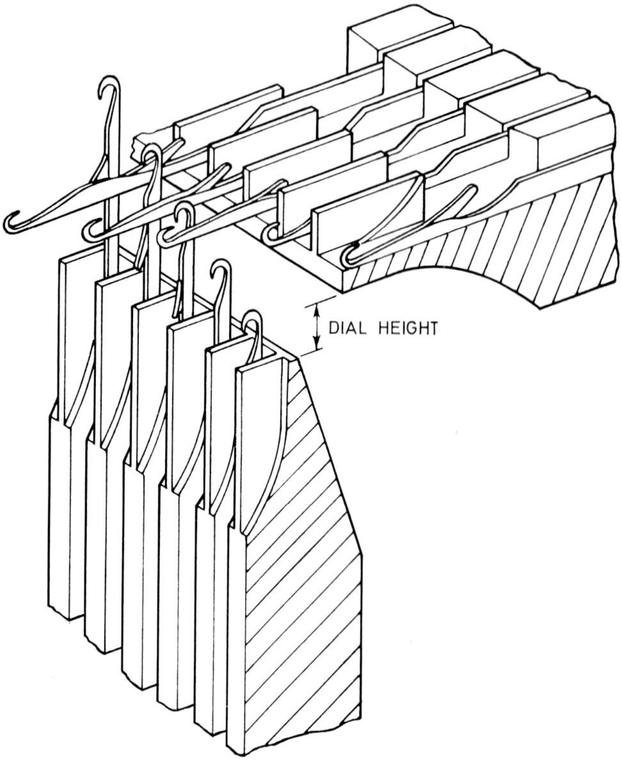

Fig. 4.7. The definition of dial height.

this will ensure both that the yarn paths between dial and cylinder are a minimum, and also that as much as possible of the fabric weight and take-down tension is carried by the cylinder needles. Using the smallest practicable dial height enables the maximum size of loops to be drawn by the needles for a given tightness, thus reducing the strains on the yarn at knock-over.

4.8 Take-down Tension

It is generally accepted that, provided take-down tension is high enough to hold the yarn on the needles and pull the old loops clear of the needles at knock-over, it can vary between fairly wide limits without affecting either knitting performance or fabric quality. Excessive take-down tension leads to breakage of the yarn in the old loops, and should therefore be avoided. Further, high take-down tensions can only be achieved by means of high roller nip pressure and this could lead to serious edge creasing if a split roller is not used. If positive feed is not used, as in Jacquard patterned fabrics, loop size can be slightly affected by take-down tension. Fortunately this is not significant as take-down tension can vary considerably around the circumference of the machine.

4.9 Stretcher Board Width

The primary purpose of the stretcher board, as its name suggests, is to spread the tube of fabric as it approaches the take-down roller so that it enters the nip and is subsequently batched without the formation of longitudinal creases. It is found that stretcher board width needs to be adjusted to accommodate different fabric structures, and this is a matter which is usually fully understood by the skilled mechanic. There is also some evidence that the setting of a stretcher board can to some extent affect the distribution of take-down tension around the circumference of the machine. However, this latter effect is insufficient to affect knitting performance.

Whenever the stretcher board is reset it will also be necessary to reset the split take-down roller. One end of the take-down rollers is turned down to a lower diameter and fitted with a series of collars which may be splined to the roller, as shown in Fig. 4.8. The positions of the collars on the shaft are adjusted so that the edge of the fabric, where it is folded, does not come under pressure but runs through the gap between two of the collars.

Normally, it is sufficient to leave 7 mm of the fabric free.

It is customary to remove two or three needles out of the machine so that the ladder thus produced lies exactly on the crease formed at the other edge of the fabric, thus providing a guide for the subsequent slitting operation. In some cases slitting is performed on the knitting machine so that edge marking in this way is not necessary.

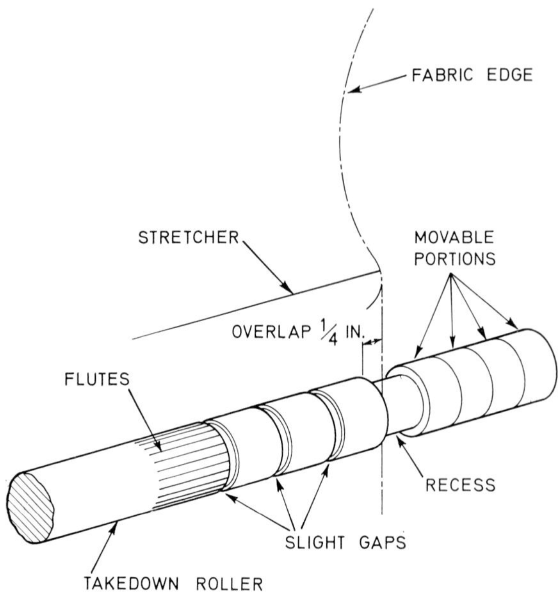

Fig. 4.8. The arrangement of a split take-down roller.

4.10 Machine Speed

It is common practice for knitters to vary their machine speeds according to the material or pattern which is being knitted. In the case of pattern the practice is reasonably consistent. The plain fabric designs are knitted at high speed, usually at or above the manufacturer's recommended speed. Jacquard designs, recognised as being more difficult to knit, are frequently run at lower speeds, thus reducing machine production rates and increasing the cost. The situation with regard to material is more ambiguous. Some knitters run wool at lower speeds than textured yarns, others run textured yarns at lower speeds than wool, both presenting plausible reasons for doing so.

All knitting machines are, of course, designed to operate at a certain speed, and this speed is recommended to users. The effect of machine speed has not been systematically studied in very great depth. What evidence there is suggests that knitting performance deteriorates slightly at speeds both above and below an optimum speed. The effect, however, does not appear to be very great, and the quantities of fabric knitted were insufficient to make any assessments of the effect of speed on fabric quality. Likewise the effect of speed on machine wear, particularly on needle butts, has not been studied.

The deterioration of knitting performance with increase of machine speed is fairly readily explained. At higher speeds the centrifugal force acting on the needles, particularly the cylinder needles, lead to a certain loss of control over their motion. This effect is widely known as "needle fling" and its importance in knitting performance is recognised.

It was mentioned in §4.4 that non-linear cams, although reducing the number of feeders on the machine, could be run at higher speed to restore the level of fabric production. However, this problem of needle fling would be of considerable importance in this respect and the problem has not yet been resolved.

The reasons for deterioration of knitting performance at lower speeds are more obscure. It may be that the lesser forces and accelerations generated at lower speeds are inadequate to form the knitted loops adequately, or the timing of the knitting elements may be subtly changed in some way hard to define, but analogous to the labouring of an internal combustion engine running at low speed.

5

MACHINE MAINTENANCE AND SETTING-UP PROCEDURES

5.1 Maintenance

Adequate machine maintenance is often one of the most difficult functions to be set up and sustained in the production unit. This is partly because it is difficult to determine what level of investment in maintenance gives the optimum return in terms of both product quality and machine efficiency. Also, responsibility for maintenance occurs at many different levels; the decision to replace a broken or damaged needle is taken at shop floor level as soon as the fault is observed; the decision to re-needle a machine involves a protracted machine stoppage, and would probably be made by the production manager.

It is convenient to consider maintenance as being in three categories. The first is general machine maintenance which is, or should be, integrated with the general plant overhauling and replacement programme. This would include such matters as the remachining or replacing of worn parts, either on a preventative basis or on breakdown. Maintenance of this type is relatively infrequent and its effect on performance is long term.

The second category of maintenance procedures is probably the most important. This involves the maintenance of the variable machine settings at the correct levels, and these should be carefully checked each time a change of product is made on a machine. These checks and any necessary corrections could normally be made by a mechanic. Any changes in these settings would be detectable, either by the knitter or during fabric inspection, so that this type of fault would be detected at most within a few days and might be described as medium term in its effects.

The third category of maintenance is combined in the duties of the knitter. It is closely related to housekeeping and its importance is often overlooked for this reason. This is concerned with general cleanliness, ensuring that thread guides are clear of dirt and fly, seeing that stop motions operate freely. Other aspects are concerned with maintaining correct machine speed, lubrication, and, as already mentioned, replacement of faulty needles. Neglect of these matters can lead to machine stoppages, fabric holes, and other fabric faults and is immediately apparent. This is thus a matter of short term maintenance.

Within each category of maintenance there are some tasks which are more important than others and some are easier to organise than others. Successful maintenance lies in applying the correct priorities and incentives to get a balanced achievement.

5.1.1 *Long Term Maintenance Planning*

Most of the long term deterioration of machinery is caused by deficiencies in lubrication systems. Most deterioration of this kind, due to mechanical wear, must be accepted as being inevitable. It can be considerably accelerated by neglect of lubrication, and it is therefore vital that lubrication routines be rigidly adhered to. In the last resort, most of the gearing drives, excepting those connecting dial and cylinder but including the take-down mechanism, may be considered to be replaceable—at a price. The most important wear damage is associated with the knitting elements. This does not apply to the needles, as these can be readily replaced and this should be done regularly. Likewise wear of the various cam plates can be overcome by either replacing or regrinding the cams. The most important wear probably occurs on the tricks and verges of both dial and cylinder. This can lead to mis-location of the needles and, if positive feed is not used, as in Jacquard designs, to variation in loop size and course length. Building up and regrinding the worn elements, although possible, is expensive and a lengthy operation and, in extreme cases, it may be necessary to install positive feed on such a machine and restrict its use to plain fabrics. Such drastic curtailment of a machine's usefulness is normally encountered only on old machines, but wear of this type could be observed during routine needle changes and the mechanic should be instructed to look at the tricks and verges carefully at these times.

Other types of wear are not associated with lack of lubrication. Wear or damage to yarn guides, stop motions, feeders, and take-down rollers can occur almost imperceptibly and in many cases can be difficult to detect. Likewise, inspection of such units is not normally included in the duties of either knitter or mechanic, so that maintenance and replacement can easily go by default. Yarn tensioners and top stop motions are particularly difficult to inspect, although the latter in particular can receive much hard treatment in the process of re-threading. All stop motions, those monitoring the needle action as well as those monitoring the yarn, should be regularly checked to see that they are still operational. Pivot points should be cleared and repairs made if necessary, and faulty units replaced.

Cracked, chipped or worn yarn guides should be replaced. Yarn guides are made with inserts of either porcelain or a synthetic ceramic. Porcelain or even plain metal guides give good service with wool yarns. These would be badly worn by synthetic fibres and for these synthetic ceramic guides should be used. Many machines are fitted with feeders having two feeder holes lined with ceramic or porcelain respectively. If a machine has been run with synthetic fibres, any porcelain or plain metal yarn guides should be checked for damage or "nicking" before running wool.

It is possible also that the fluting of the take-down rollers might wear differentially. This could be expected to occur only if slippage of the fabric had taken place over a period of time. To overcome this, it is usually sufficient to regrind and re-cut the rollers, but in extreme cases replacement may be necessary. In either case, an unplanned machine stoppage can occur if the incidence of wear remains undetected. Normally, wear is fairly easily detected by the skilled observer provided only that he has been asked to report and—probably more important—if his reports are acted upon.

Probably the most insidious type of wear, adversely affecting both fabric and needle life, is that of cams. Further, since they are not immediately visible they can be in use for protracted periods without attention and positive action has to be taken if they are to be inspected. Once found, worn cams should be reground or replaced and star wheels re-set.

5.1.2 *Medium Term Maintenance*

Most of the variations due to medium-term effects can occur as a result of machine adjustments. The adjustments most frequently made to a machine are cam settings, needle selections, dial height, positive feed or yarn tension control, stretcher board width, and take-down tension. The nature of these adjustments will be discussed in §5.2.

Of these, needle selection, cam setting, and positive feed or yarn tension control are completely determined by the fabric design. However, the opportunity should be taken to check the functioning of the selection system and of the yarn supply system. In re-setting dial height, care should be taken in ensuring that it is the same all around the machine, otherwise fabric quality and number of holes in the fabric will be adversely affected. Instruments exist for the measurement of yarn tension and speed and of dial height, so the control of these variables is not particularly difficult.

Control of settings of the stretcher board and the take-down rollers is not easy since measurement of the take-down tension is difficult. Precautions must be taken to ensure that the stretcher board is not damaged and, if supported, is centrally located and not tilted. Take-down rollers are based on either the two-roller or three-roller system. By inserting metal strips of the same thickness at intervals along the nip, it is possible to check by rotating the rollers that they are straight, parallel, and evenly worn. If wear has occurred on one part of the roller, the metal strip will be easily removed from that part.

Deviations from parallelism can be corrected by re-setting the roller bearings relative to one another. Bent or unevenly worn rollers can be reground. Alternatively, it may be deemed easier and cheaper to replace the whole take-down roller assembly.

Nip pressure should be checked at each end of the take-down roller by means of a spring balance acting on a harness passed around the loaded roller. It is important that the nip pressure should be as low as possible to minimize creasing and that it is also uniform across the take-down. It may be useful to build up a record of pressures that are suitable for different fabrics.

5.1.3 *Short-term Maintenance*

Much of this vitally important component of maintenance has already been discussed. It involves cleaning of the knitting zone, feeders, thread guides, and proper use of the machine. Most of this type of activity is considered as part of the duties of a knitter, but it should be more properly considered as maintenance, rather than operation of the machine.

The proper maintenance of the expensive, but highly productive, knitting machine is a complex operation and requires the joint efforts of manager, mechanic, and knitter, to obtain best results.

5.2 Setting-up Requirements

Setting-up a double jersey machine is a skilled and complex operation, this being reflected in the remuneration of those who do it. It is sound economics therefore to make the task as easy and rapid as possible, particularly if maintenance checks are to be included in the setting-up procedure. The mechanic's requirements may be divided into two categories, equipment and information, and these will vary according to the fabric for which the machine is being set-up.

5.2.1 *Equipment Required*

Apart from his conventional tools, a knitting mechanic will require a yarn tension meter, a yarn speed meter, and a course length gauge, in order to set a fabric to the required tightness. A dial height gauge will be required to set and level the machine dial, and a set of feeler gauges and a spring balance to check the take-down rollers. Additionally, if a patterned fabric is to be produced, he must be supplied with the complete patterning system, i.e. drums, keys and wheels, ready for mounting on the machine. There must be no question of assembling pattern units while the machine is stopped.

5.2.2 *Information Required*

The information available to the mechanic may vary from fabric to fabric, but it is essential that it be adequate for his purpose. For example, for a plain fabric he should be given the yarn input speeds for a certain machine speed, or the course lengths. (In many cases it may also be possible to give approximate cam settings to assist in the setting-up process). For Jacquard fabrics, fabric weight is the only information usually available, and cam setting is usually by trial and error in order to achieve the required weight.

The mechanic will also need to have some information on the structure of the previous fabric on the machine, since the relationships between the old and new yarn counts and fabric tightness may modify to some extent the sequence of operations whereby he resets the machine. For example, although it is normal to change the yarn as the first operation, if the new yarn were appreciably coarser than the old it may be necessary to raise the dial before the new yarn can safely be run in.

5.3 Setting-up Procedures

The procedure for setting-up a machine may be considered under two aspects; the techniques whereby the various controls and adjustments are made, and the order in which the adjustments are made. As indicated above, this latter aspect depends to a certain extent on both the old and the new fabric. It also depends to some extent on the machine, particularly the presence or otherwise of a positive feed system. If a machine is equipped with some form of constant tension device, then it would be unusual to change the yarn input tension. Generally speaking, however, it will be found necessary to re-adjust a positive feed system for every plain fabric that is knitted. Positive feed systems are not suitable for Jacquard patterns (including blister fabrics) and so would be taken out of service for these fabrics.

5.3.1 *Setting-up Plain Structures*

Under normal conditions the recommended sequence for setting-up for a plain structure would be:

- A Change yarn
- B Gating and needle timing, feeder positions
- C Dial height
- D Needle selection
- E Stitch cams
- F Positive feed (if available)
- G Take-down
- H Stretcher board
- I Check fabric appearance and quality
- J Check yarn input tension and yarn speed

This sequence would be used if the yarn change were at the same count or to a finer count. If the change was to a coarser yarn count, then it is strongly recommended that the dial height be increased before the new yarn is run into the machine.

5.3.2 *Setting-up for Jacquard Fabrics*

A Rib gate, needle timing (synchronized), feeder positions
B Change yarn
C Dial height
D Select dial needles
E Select cylinder needles (change patterning elements)
F Stitch cams
G Check pattern and quality
H Take-down
I Stretcher board
J Yarn tension
K Machine speed

If the change is from a plain fabric it is imperative that the positive feed be disconnected first. Again, if the change is to a coarser yarn, raising the dial should precede the change of yarn. It is frequently found necessary to run knitting machines at speeds lower than the recommended speeds when Jacquard patterns are knitted.

5.3.3 *Setting-up for Blister Fabrics*

Essentially blister fabrics resemble Jacquard pattern fabrics but they are more complex and it is easier to regard them as a separate class of fabric.

A Rib gate, needle timing (synchronized) and feeder positions
B Change yarn
C Dial height
D Fit blister cams as required
E Select dial needles

F Select cylinder needles (change patterning elements)
G Stitch cams. Adjust dial height if required
H Rethread all feeders knitting on dial only (vertical guide)
I Fit float detectors if required
J Check needles for clearing and loading-up
K Take-down (high tension required)
L Stretcher board
M Check fabric quality and pattern
N Check yarn input tension
O Check machine speed

Again it is assumed that any positive feed would be disengaged.

There are two or three items here of special interest since they occur only on fabrics of this type. The first is the use of blister cams on the dial. These cams are used where the pattern calls for knitting to take place on the cylinder only. Under these circumstances there is a danger, due to the absence of sinkers, that the old loops on the cylinder needles would rise with the needles instead of being formed behind the latch. This would cause loops to accumulate on the needles with severe consequences. The remedy is to bring the dial needles some way forward of the rest position, but not far enough to take the dial loop behind the latch. In this position they act as a type of sinker arrangement, preventing the cylinder loops from rising. Once the cylinder needles are fully raised, the dial needles are withdrawn so that yarn is not fed into them. Apart from this much earlier retraction, the process is very similar to tucking.

Where feeders are knitting on the dial only, the feeder should be re-threaded so that the yarn is running in the vertical guide. This is to ensure that the yarn is fed right into the dial needles, the usual horizontal guide not being low enough for this purpose.

Float detectors may be fitted, if thought necessary, behind cylinder feeds and in line with the stitch cams. They would be used in order to stop the machine if the location of larger floats was faulty.

5.3.4 *General Aspects of Setting-up*

Knitting machines are intricate and relatively fragile units, and during setting-up they must be treated with care. They are normally fitted with three speed controls (apart from manual rotation). These are "run", "inch", and "crawl", and these are all available at multiple points around the machine. Once started, the "run" speed requires the use of the stop button to stop the machine. The "crawl" button causes the machine to turn at a very slow speed and the "inch" button at an intermediate speed, but only while the button is held depressed.

The "inch" or "crawl" button should always be used for running in new yarn, making new needle selections, and checking such settings as dial height and feeder position. This is to ensure that the machine is always under the control of the mechanic and risk of damage is minimized.

The setting of stitch cams and the checking of loop clearing can best be done with the machine running slowly. Setting of positive feed speeds and yarn tension can only be done when the machine is running at production speed. Adjustment of take-down tension and stretcher board width is probably best accomplished after the machine has run at full speed for some time and equilibrium conditions have been obtained.

It is, however, of vital importance that the final check on fabric quality, yarn tension and speeds, and machine performance should be made after the machine has been running at production speed for some time. It is well known that it takes an appreciable period of time for the lubricants to attain equilibrium temperature, and the final running speed of a machine is normally rather higher than its initial speed.

5.4 Setting-up Techniques

The procedures listed in §5.3 showed the sequence for setting-up knitting machines. Some of the steps listed will now be described in more detail. Others have been discussed in Chapter 4. Much of this information will be given in the machinery manufacturers' handbooks and small differences in numerical data may be found. Nevertheless, the following is substantially accurate and some of the comments and observations may be of additional value.

5.4.1 *Gating*

For interlock fabrics, and for most types of plain fabric, interlock gating is used. In this type of gating, dial and cylinder needles lie opposite each other so that if both were raised together they would collide.

Rib gating must be used for all patterned fabrics where dial and cylinder needles are independently selected. Rib gating consists of arranging the dial and cylinder so that the dial needles lie midway between the cylinder needles, making impact impossible, as shown in Fig. 4.7. There is at the present time a growing tendency to knit all fabrics, plain as well as patterned, on rib gating. This eliminates the risks of needle breakage on pattern change and also reduces setting-up time.

5.4.2 *Timing*

Patterned fabrics are normally knitted using synchronised timing (i.e. with the dial needles knocking over half a needle after the cylinder needles) but Jacquard patterns can sometimes be knitted with perhaps two or three needles delay thus providing a slightly tighter fabric. (See Fig. 4.6). Plain fabrics can be knitted with up to nine needles delay, but delays of more than six needles are exceptional.

Timing is adjusted by loosening off the cam plate locking screws and then rotating them the required amount, subsequently tightening the locking screws. Depending on the machine, it may be either the dial cam plates or cylinder

Fig. 5.1. Using a dial height gauge. Courtesy Hosiery and Allied Trades Research Association.

cam boxes that are moved, but whichever is the case they move as a unit and there is no question of re-setting individual cams to the new timing.

5.4.3 *Feeder Position*

Feeders should be positioned at optimum distance from the dial and cylinder needles, and also relative to the knock-over position. These positions are normally given by the machinery manufacturer and are usually in the range 0·015-0·018 in. (400–450 microns) from both sets of needles.

5.4.4 *Dial Height*

The dial height should be such that the fabric can pass freely between the dial and cylinder blocks. Normally, dial height is not changed except in the case of a change to or from abnormally fine or coarse count yarns or when blister fabrics are being knitted. Adjustment of dial height should never be used as a means of controlling fabric quality. Dial height is set by a dial height gauge which measures the distance between the cylinder verge and the dial needles (Fig. 5·1). However, it is the separation between dial and cylinder which is of importance and this differs from the height measured by the gauge and must be allowed for. When lowering the dial it is good practice to lower too much and then to raise it to the required height. The dial must never be set so low that a needle from that machine cannot be passed between the gap. This is a safety measure to ensure that if a needle should break the debris can be disposed of without damaging the machine.

5.4.5 *Yarn Input Tension*

The yarn input tension should be in the range 2–3 g. This is achieved in one of three ways, depending on the installation in use. If positive feed is being used, then once the yarn input speed has been set the yarn tension is adjusted by setting the stitch cams. If a yarn tension feed system is used, then this, once set, should require the minimum of attention. Details of the methods of setting these types of feed vary, and information is given by the suppliers. If driven feed systems are not used the yarn tension control is achieved by means of disc or cymbal tensioners in the yarn path, and the weighting of these is varied as required.

Fig. 5.2. A yarn tension meter in use.

5.4.6 *Take-down Tension*

Take-down tension will require adjustment wherever a change is made to fabric structure. This has been discussed in §4.8. The take-down tension should be as low as possible, consistent with achieving satisfactory knock-over and, as indicated in §5.3.3, it will be found that higher take-down tensions are required for blister fabrics.

5.4.7 *Fabric Take-up*

This usually takes the form of a batching mechanism driven through an adjustable ratchet system in conjunction with the take-down rollers. This again is a non-critical setting, but it is necessary to strike a balance between two extremes. On the one hand, loose rolling reduces the risk of edge creasing and introduces the possibility of fabric relaxation, which may or may not be desired, and produces a low weight roll. Tight rolling preserves fabric dimensions and increases the roll weight but may lead to edge creasing.

Some machines allow the fabric to drop into a bin after passing through the take-down rollers. This allows more fabric relaxation and reduces edge creasing but introduces the possibility of random creasing.

6

PIECE TREATMENTS

The objective of any finishing operation is to produce a fabric with an acceptable hand, appearance and performance, particularly with regard to stability.

Whereas many other types of wool fabrics are wet finished in order to remove impurities and acquire the desired softness of hand, the requirement of a firm, crisp hand in double jersey fabrics prohibits, in most cases, the application of normal wet finishing procedures. A dry finishing procedure is therefore usually adopted except in those cases where the requirements of coloration, special finish effects and, in some cases, washing performance, necessitate the use of a wet process.

Another difficulty in finishing double jersey wool fabrics is the high inherent relaxation shrinkage. An area shrinkage of 20 per cent on soaking in water is not uncommon for double jersey fabrics, and this can increase to up to 40 per cent if the fabric is given a mild agitation in water. A more severe agitation results in felting or irreversible relative fibre movement.

Most wool double jersey fabrics are sold under a "Dry Clean Only" label, so that the only stability requirements are that the fabric should not shrink excessively in tailoring and dry cleaning. Shrinkages of not more than 2–3 per cent in any one direction are tolerable and can be readily achieved by a simple dry finishing procedure involving, at most, steam relaxation followed by blowing or semi-decating. In some cases semi-decating alone is adequate.

Washable finishes imply the control of both relaxation and felting shrinkage, the former being achieved by a combination of a partial relaxation and a partial setting treatment and the latter by an anti-felt treatment applied at the top or fabric stage.

6.1 Slitting, Inspection and Rolling

If a fabric is stored in a folded state for a prolonged period, local relaxation occurs at the fold and this may show up as a mark down the piece which is difficult to remove. This is particularly critical for fabrics which are piece dyed at open width. A preferred procedure is to slit and cuttle at the knitting machine or, if it is necessary to roll at this stage, the fabric may be slit by a knife inserted before the take-down rollers. The width of the take-down rollers should be adjusted to avoid pressure on the crease itself. If it should be necessary to wet process at open width fabrics which have been held in the roll for a considerable time, some success in removal of the centre crease can sometimes be obtained by a partial steam relaxation followed by semi-decating at high wrapper tension before wet treatment.

During inspection, faults should be marked and records kept of the fabric dimensions, in order to predict future dimensional changes in finishing.

Generally faults which are impracticable to mend are "stringed" and discounts made to the customer accordingly.

6.2 Mending

Wool is an expensive raw material so that it is necessary, where possible, to mend any defects in the fabric. If any part of the finishing proccss necessitates processing in tubular form, the fabric must be mended in this form. Mending is a skilled operation calling for a variety of techniques for best results.

6.2.1 *Slubs and Burrs*

Very short slubs caused by fly can be removed by picking with tweezers. Larger slubs are not usually mended unless it is considered absolutely necessary to do so, in which case the defective yarn is removed and replaced by a technique similar to that used for the mending of holes.

Burrs (vegetable contaminants) are removed by picking out with tweezers.

6.2.2 *Knots and Loose Ends*

Knots and loose ends are cut to between 0·25 and 0·5 in. (6–12mm) and pushed into the body of the fabric, taking care to avoid the formation of holes by cutting too close to the fabric.

6.2.3 *Dropped Stitches and Small Holes*

Dropped stitches are picked up with the aid of a latch needle or turning hook and prevented from dropping again by looping the dropped stitch and adjacent stitches with a short length of yarn.

Small holes are mended by interlacing new yarn with about four stitches on each side of the hole and then interlacing the new yarn with existing stitches using a latch needle or turning hook.

6.3 Dry Finishing

The usual dry finishing procedure consists of an open steam relaxation followed by semi-decating or blowing.

Steam relaxation machines usually consist of a perforated lattice onto which the cloth is fed. Steam is blown through the lattice to relax the fabric, followed by drying with hot air. Some machines are equipped with vibrating systems to give further relaxation but for "Dry Clean Only" merchandise, this is probably not essential. If a steam relaxation machine of the type described is not available, relaxation may be achieved by a steam box at the front end of the stenter, setting the rails so that the fabric enters the drying

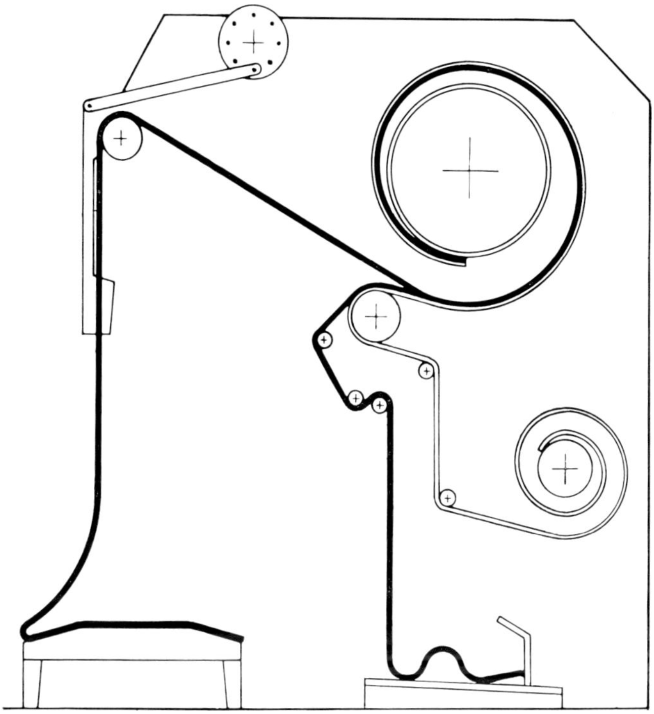

Fig. 6.1. Sectional view of a semi-decating (steam-blowing) machine. The fabric is fed in along the lower route shown and removed along the upper route.

zone with a slight sag and adjusting the overfeed so that the undulations in the fabric just disappear as the fabric leaves the steaming zone.

It is convenient to remove the fabric from the relaxation process in the form of a batch and feed into the blowing machine by contact unrolling from the batch. An adequate degree of stability can usually be obtained by continuing the steaming for one minute after initial steam penetration across the width of the roll. The fabric is cooled on the

Fig. 6.2. Piece dyeing of wool double jersey by the cold pad-batch process.

roll either by suction or by blowing air through the roll of fabric. The latter procedure is preferable if lustring or flattening of the fabric is not required. Low wrapper tensions and the use of raised "molleton" wrappers also tend to decrease lustring and flattening.

After blowing, the fabric is re-batched. Contact unwinding onto cardboard tubes is often preferred to cuttling and rewinding, as this reduces the possibility of imposing strain on the fabric.

During finishing, it is sometimes advisable to add sewing lubricants in order to prevent needle damage in sewing. This is particularly important in cases where fabrics are dry cleaned to remove oil stains. A limited number of oil stains can be removed from wool by spot cleaning and a full dry clean is necessary only for pale shades containing a large number of oil spots.

6.4 Piece Dyeing

Piece dyeing has obvious advantages from the point of view of improved stock control, shorter delivery times, and the superior knitting performance of an ecru yarn. However, double jersey fabrics are mostly knitted from coloured yarns, even when solid shades are required. Piece dyeing has not been used to any great extent for double jersey fabrics because of the possibility of producing an unsatisfactory fabric with regard to stitch distortion, surface fuzzing, formation of longitudinal creases, excessive lustring, flattening and loss of the desirable crispness of hand. There is also a problem of high shrinkage during processing resulting from the high relaxation shrinkage present in this type of structure.

Recent technological developments, particularly in the handling of double jersey fabrics during dyeing, have overcome many of these disadvantages. Generally, fabrics can be satisfactorily dyed in open width by procedures such as cold pad batch or beam dyeing. Even winch dyeing is acceptable in some cases where fabrics have been produced from yarn which has been subjected to a satisfactory anti-felt treatment.

6.4.1 *Cold Pad Batch Dyeing*

This consists of padding the fabric in a solution of dye, surfactant, thickener, sodium bisulphite and urea, followed by batching onto a roll, wrapping with polythene, storing from 4–24 hours and washing off.

A typical pad recipe (based on 100 per cent pick-up) is:

300 gl^{-1} Urea
10 gl^{-1} Dispersol VP (I.C.I.)
5 gl^{-1} Guaranate AP5
1 gl^{-1} Acetic acid
x gl^{-1} Dye
10–20 gl^{-1} Sodium bisulphite (the higher concentration being used for dye concentrations greater than 10 gl^{-1}).

This procedure has the considerable advantage of minimum stitch distortion, no surface fuzzing and very little deterioration in hand. It is, however, essential that the fabric be batched at minimum tension in order to avoid excessive flattening and the production of moiré effects.

Another essential is that the fabric should be washed off at open width and that no length tension be imposed during washing off. A beam washer is an ideal machine for this purpose.

The width shrinkage during processing is approximately 10 per cent but this depends on fabric structure and pre-treatment. Semi-decating the fabric before padding reduces the width shrinkage to 3–5 per cent. If problems of excessive width shrinkage during processing occur, it may be necessary to modify the fabric structure. For example, for a Punto-di-Roma fabric, such a modification would consist of decreasing the tightness or increasing the run-in ratio or both. The effect of fabric structure on dimensions is discussed in Chapter 7.

6.4.2 *Beam Dyeing*

If correctly carried out, this procedure, like pad batch dyeing, is capable of maintaining stitch clarity and freedom from surface fuzzing. Beam dyed fabrics are slightly limper than pad batch dyed fabrics, but the difference in hand is not great. Perhaps, more than with any other process, the potential advantages of beam dyeing can be fully realised only if adequate precautions are taken.

A typical procedure is:

Slit
Open
Mend
Pre-relax
Wind on to beam

Fig. 6.3. A beam-dyeing machine. Courtesy Platt International Ltd.

Scour } on beam
Dye } on beam
Shrink resist } on beam
Hydroextract or squeeze
Dry
Semi-decate (if desired).

If the fabric has a high potential width relaxation, excessive shrinkage can occur on the beam during dyeing, resulting in marks along the length direction of the fabric which are extremely difficult to remove after dyeing. High length relaxation on the beam can cause excessive flattening, lustring and the development of moiré effects, resulting in a totally unacceptable handle and appearance. These problems may be overcome by a partial relaxation followed by beaming at low tension.

The required amount of relaxation may be accomplished by open steaming or by padding with water and drying to the relaxed dimensions. If dry cleaning is used to remove oil stains, the technique can be modified to act as a relaxation procedure. Very little shrinkage takes place when wool jersey is dry cleaned in a pure solvent system, but the procedure is capable of permitting high relaxation shrinkages if a charged system of dry-cleaning is used. This aspect is discussed in detail later in §6.5.5. A danger of this procedure is that stitch distortions frequently occur during the operation. These distortions can be avoided by blowing or semi-decating the fabric before cleaning in a charged system.

In beaming, it is essential that the layers of fabric are superimposed exactly on each other, that the fabric is wound free from distortions and creases and that no length strain is imposed during winding.

The ideal system for winding fabrics on the beam consists of a stenter feed to a peripheral batching arrangement. Successful results may be obtained by winding on a single pass stenter using a steam box at the feed end in order to achieve the required pre-relaxation.

As the fabric is wound on to the beam at low tension it can move and crease during dyeing unless means are adopted to counteract this tendency. A successful procedure is to use a wrapper consisting of an elastic fabric or a normal fabric wound at high length tension. This fabric must be sufficiently wide to enable it to be tied to the beam at the outside edges of the roll. Three or four wraps are sufficient to prevent movement.

The operations of scouring, dyeing and washing-off are carried out consecutively in the beam dyeing machine. In order to avoid excessive flattening, the bath should not be drained until the completion of the wet processing sequence.

Most beam dyeing machines are suitable for dyeing wool providing that the pump gives adequate circulation. Probably the most successful machines for the avoidance of excessive flattening utilise air injection.

Unless the fabric has been previously dry cleaned, scouring on the beam is necessary to ensure even treatment.

The usual procedure is to scour for 15 minutes at 40°C in a solution containing 1 gl^{-1} of a non-ionic detergent and 1 gl^{-1} sodium bicarbonate. After the scour, the material is rinsed with water for 5–10 minutes.

The dyeing procedure used depends on the class of dyestuff being applied. For reactive dyes, the fabric is adjusted to pH5 by adding acetic acid (approximately 1 per cent o.w.f.*) and the dyeing assistant (e.g. 1 per cent o.w.f. Albegal B (CIBA-Geigy)). The liquor is circulated for 10–15 minutes and the pH checked. The dyestuff is introduced and circulation continued for a further 10–15 minutes prior to raising the temperature at 1°C per minute to 70°C, held for 20 minutes at this temperature and then raised to 105°C, at 1°C per minute and held at this temperature for about 20 minutes, after which time the exhaustion should be complete.

The temperature is than lowered slowly (1°C per minute) to 80°C and the pH adjusted with ammonia to pH8. After 15 minutes the fabric is rinsed with warm and then with cold water.

Although the above discussion refers to the use of reactive dyes, other types of dye can obviously be used.

6.4.3 *Winch Dyeing*

Winches are by far the most numerous of all the piece dyeing machines in existence, so that it would obviously be advantageous to be able to use a winch dyeing procedure for the dyeing of wool double jersey. Problems of excessive fuzzing and felting during winch dyeing are virtually eliminated by knitting the fabric from anti-felt treated yarns but the major problem of excessive relaxation shrinkage during dyeing remains.

Due to the length strains involved in this system of processing there is a slight increase in the length of the fabric and a very pronounced decrease (in the region

* On the weight of fabric.

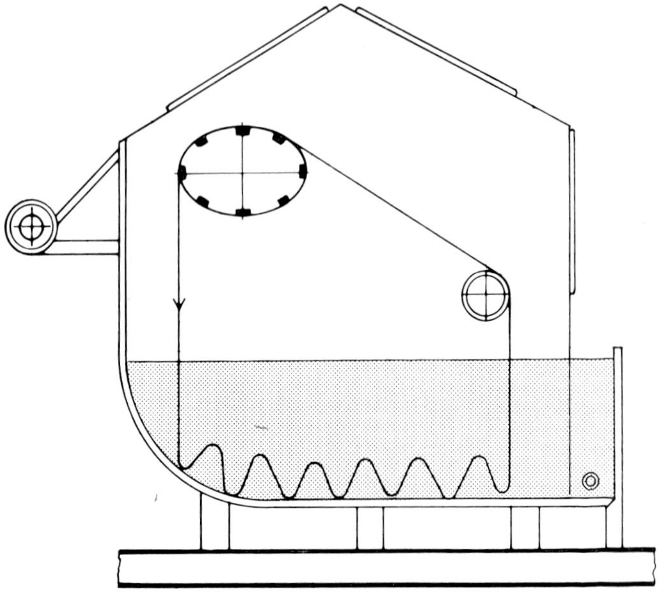

Fig. 6.4. Sectional view of a winch dyeing machine.

of 25 per cent) in fabric width during dyeing. This means that a fabric 66 inches (168 cm) wide would shrink to approximately 50 inches (127 cm) after dyeing. Such a width would probably be unacceptable to the maker-up.

It would of course be possible to achieve a significantly greater width by stretching in the width direction during drying, but this would result in high shrinkages during making up and use, unless the imposed strains were stabilised by a setting procedure. Autoclave setting after drying is an obvious possibility but only about 50 per cent of the imposed strains can be stabilised by this procedure, the remaining 50 per cent returning as shrinkage during prolonged washing. The stability to tailoring and dry cleaning would of course be higher than the stability to washing.

It is generally recognised that a maximum shrinkage tolerance in washing a double jersey garment is 4 per cent in the width direction, so that, assuming a 50 per cent setting efficiency, the maximum amount of width extension that could be imposed in drying would be 8 per cent, giving a fabric width of 54 in. (137 cm), which might just be acceptable for dress fabrics. The shrinkage occurring during winch dyeing is dependent also on the fabric structure, tightness and run-in ratio so that acceptable fabric widths could be obtained for some fabrics processed by this route, whereas others would give unacceptable finished widths.

An alternative procedure is to autoclave set before dyeing. This reduces the width shrinkage in dyeing by about 50 per cent, e.g. from 25 per cent to 12 per cent, so that a stable finished width of 58 in. (147 cm) could be obtained. The major problem of using this procedure is that the fabric must be slit before setting and dyeing, which could result in the development of winch marks. To avoid these, it might be necessary to sew up again into tubular form before dyeing.

In general, the most satisfactory procedures from the point of view of preservation of handle, stitch clarity, and general appearance of the dyed fabric are cold pad batch and beam dyeing. The change in handle and appearance coupled with a narrow fabric width renders winch dyeing a less desirable route.

6.5 Washable Finishes

As mentioned previously, in order to produce washable double jersey fabrics, both relaxation and felting shrinkages must be controlled. Low relaxation shrinkages may be obtained by a combination of a partial relaxation and a partial setting procedure, and low felting shrinkages by the use of shrink-resistant yarns or by application of an anti-felt treatment to the fabric.

Possible routes are:

(i) Autoclave setting of fabrics produced from shrink resistant yarns.
(ii) Beam dyeing of fabrics produced from shrink resistant yarns.
(iii) Beam dyeing and application of an anti-felt treatment while the fabric is on the beam.
(iv) Application of aqueous resin emulsions.
(v) Application of solvent-soluble resins.

6.5.1 *Autoclave Setting of Fabrics Produced from Shrink Resistant Yarns*

This treatment can be applied to fabrics knitted from yarn which is spun from wool which has been given a satisfactory oxidative or resin anti-felt treatment. After partial relaxation by one of the methods outlined under Section 6.4. the fabric is wound onto a perforated beam and treated in an autoclave with steam for 5 min. at a pressure of 1 atmosphere (15 lb in.$^{-2}$).

The method of treatment depends to some extent on the type of autoclave equipment available. In a pressure decator of the type used in the finishing of woven fabrics, the fabric is wrapped on the perforated roll between layers of cotton molleton wrapper, using low wrapper tension in order to prevent excessive flattening. For other types of autoclave which do not employ a wrapper, it is sufficient to wind the cloth directly onto a perforated roll covered with a wrapper to prevent marking and then to cover with a few layers of wrapper, which are tied at the ends, to prevent the fabric coming off the roll and to minimize condensation. In either case the roll must be supported at the ends during the autoclave treatment.

In autoclaves where steam is blown through the centre of the roll and then vented to the outside atmosphere, the vent is closed and timing commenced when the required pressure is reached. In the cases where the autoclave acts merely as a closed container, a double vacuum cycle must be used to obtain adequate penetration through the roll.

6.5.2 *Beam Dyeing of Fabric Produced from Shrink-Resistant Yarns*

As beam dyeing produces a high level of set, beam dyed fabrics exhibit low relaxation shrinkage, so that, if these are knitted from shrink-resistant yarns, no further treatment is necessary in order to obtain a washable fabric. Care must, however, be taken in drying in order to avoid the imposition of strains.

Before drying, as much surplus water as possible should be removed. This can be done either by an open width mangle or a centrifuge.

The centrifuge technique is probably the more effective with regard to the removal of surplus water, but care must be taken in the subsequent opening of the fabric in order to avoid the imposition of strains. Scutches should not be used for this purpose.

A suction drum dryer is probably one of the best machines for the drying of open width knitted fabrics free from strain, but stenter drying is more usual. The stenter should preferably be equipped with a rail, lattice or tape to support the fabric, particularly at the feed end. A satisfactory procedure is to set the stenter at 1 in. (2·5 cm) over wet width, adjusting the overfeed so that the fabric does not sag appreciably.

6.5.3 *Beam Dyeing and Shrink-resist Treatment on Beam*

An alternative to the knitting of fabrics from shrink-resistant yarns is to apply an anti-felt treatment after dyeing, while the fabric is on the beam. The procedure consists of adjusting the fabric to pH5 with acetic acid and adding D.C.C.A.* (Fichlor) to give 4 per cent. D.C.C.A.*

* Dichloroisocyanuric acid (D.C.C.A.).

o.w.f. (i.e. 2·4 per cent available chlorine o.w.f.). Circulation is continued until exhaustion is complete, raising the temperature to 30°C in the later stages if necessary. This chlorination procedure, which takes approximately 45 minutes, is followed by circulation of 3 per cent sodium bisulphite solution for 15 minutes to remove the residual chlorine. Two rinses in water complete the process.

If desired, other suitable anti-felt treatments may be used as alternatives to D.C.C.A.

After treatment, the fabric must be dried free from strain by the methods outlined previously.

6.5.4 *Application of Aqueous Resin Emulsions*

The process consists of setting the fabric, pad application of resin emulsion, drying, storing for 3–4 days to allow the resin to cure, washing off the surplus chemicals and drying at low tension.

The setting process consists of steaming in an autoclave for 5 min at a pressure of 1 atmosphere (15 lb. in.$^{-2}$) by the method outlined previously (§6.5.1). Although adequate shrinkage control in washing can be obtained on most fabrics without the inclusion of autoclave setting in the processing sequence, the setting process gives a much superior after-wash appearance and fabrics treated by the route do not generally require ironing after washing.

Autoclave setting can also be used as a means of increasing fabric width and reducing fabric extensibility. Although it has been emphasized previously that stretch should not be imposed on the fabric during finishing, autoclave setting will stabilize about 50 per cent of the imposed stretch. Almost all of the imposed stretch which is not stabilized by the autoclave treatment will be removed by the subsequent wet padding operations.

After setting, the fabric is padded in a liquor containing:

3·5 per cent o.w.f.[1] Oligan 500 (CIBA-Geigy), emulsified with 50 per cent o.w.r.[2] Lissapol N (I.C.I.)[3], 0·2 per cent o.w.f.[1] Catalyst 3774 (CIBA-Geigy), and 2 per cent o.w.f.[1] potassium bicarbonate.

After padding, the fabric is dried, stored for 3–4 days to allow the resin to cure, and then washed-off. The washing-off procedure removes residual chemicals and improves the hand of the fabric.

Finally, the fabric is dried at minimum tension by the methods outlined in §6.5.2, and semi-decated.

6.5.5 *Application of Solvent Soluble Resins*

Some solvent soluble resins, notably Synthappret LKF (Bayer) can be used to control both relaxation and felting shrinkage in double jersey fabrics and garments. In order to obtain the required standards of dimensional stability it is preferable to incorporate in the process a partial setting or a partial relaxation, or both.

The resins may be applied as batch treatments in modified dry cleaning machines or continuously in continuous solvent treatment machines. The major problem with both systems of application is uneven treatment, either as a result of initial unevenness of application or of migration during drying.

In the batch application, the fabric is dry cleaned to remove any oils or waxes which would otherwise reduce the effectiveness of the treatment, dried and then circulated in a 2 per cent solution of the resin until wet-out. After spinning to 100 per cent pick-up (by weight), the solvent is dried off and recovered.

The 100 per cent pick-up is critical as, if the pick-up were too low, the resin may be applied unevenly, whereas if the

[1] o.w.f. = on weight of fabric.

[2] o.w.r. = on weight of resin.

[3] Oligan 500 is now available as a 40% emulsion. When this is used, Lissapol N is omitted.

pick-up were high, unevenness of treatment could occur as a result of migration during drying. In any case, even treatment of piecegoods in a dry cleaning machine is difficult to achieve due to entanglement of the load.

If desired, drying after the initial dry clean cycle may be omitted but, in this case, it is necessary to compensate for dilution of the resin bath by addition of new resin after every cycle or every other cycle.

If a system of drying after the initial dry clean is adopted, the dry clean cycle can be utilized as a partial relaxation process by using a charged system of cleaning, i.e. by addition to the system of an emulsion of water in perchlorethylene. The shrinkage in the charged system increases with the amount of water in the system, but no more than 10 per cent of water, emulsified with a suitable emulsifying agent, e.g. 10 gl^{-1} Milling Aids KFS (Allied Colloids) is required to give an area shrinkage in the region of 20 per cent. This should give adequate stability for most end uses. This system of relaxation results in some stitch distortion in the fabric, but this may be eliminated by semi-decating before treatment.

As the resin reacts with water, the charged system technique cannot be used in cases where the first drying is omitted.

An alternative to partial relaxation is autoclave setting, which is preferably carried out before resin treatment.

Steam tumbling or dyeing after resin treatment gives an appreciable improvement in the hand, which otherwise tends to be rather harsh.

6.6. Special Finishes

Milling and raising are examples of special finishes which can be applied to double jersey fabrics. In milling, felting does not start until most of the relaxation shrinkage has been removed. This implies that high processing shrinkages (up to 15–20 per cent in width) can occur and, although some width stretch in tentering is probably permissible for "Dry Clean Only" merchandise, this should not be excessive or dimensional instability will result.

Raising is not so effective on double jersey structures as on warp knits and single knit inlays, as the action is mostly on yarns running across the width of the fabric. This means that raising will mostly occur on the top of the loops in a double knit fabric, resulting in a low raising efficiency and an uneven pile. The unevenness of the surface can be corrected to some extent by brushing and shearing and the effectiveness of raising may be improved by slightly milling the fabric before raising.

As raising, brushing and shearing all tend to decrease the fabric width the incorporation of a further shrinkage process such as milling could result in the production of fabrics well below commercially acceptable widths.

7

FABRIC PROPERTIES

7.1 Fabric Dimensions

It is now accepted that the length and the width dimensions of a knitted fabric are largely dependent on the length of yarn in a loop, i.e.

Courses per inch, $c = \frac{k_c}{l}$

Wales per inch, $w = \frac{k_w}{l}$

Knowing the constants of proportionality, and controlling the length of yarn in a knitted loop, fabrics of predetermined dimensions can be produced. The constants, k_c and k_w, vary with the energy state of the fabric. During manufacture and processing the loops are strained and distorted and the fabric is said to be in a state of high internal energy. The extent to which the loops are allowed to assume their natural configuration, which is a state of minimum internal energy, depends on the extent of frictional restraints of the interlacing loops. The fabric thus exhibits a series of intermediate semi-stable states. This phenomenon is easily understood from a diagram (Fig. 7.1) which shows the general relationship between the fabric's internal energy level (degree of relaxation) and the geometrical constants.

Since double jersey structures are relatively complex, transfer from one energy state to another is not readily accomplished, so much so indeed that many wool double

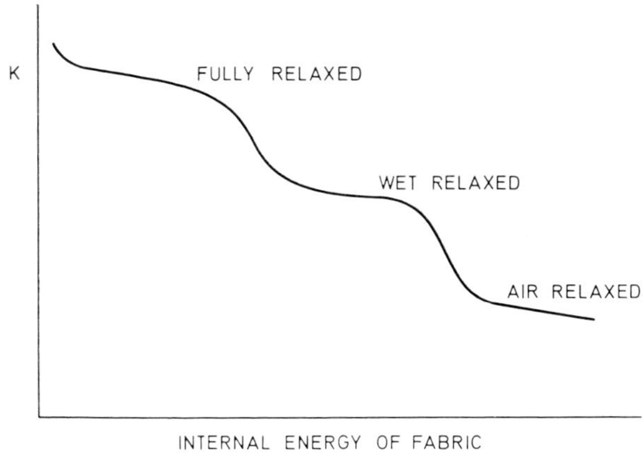

Fig. 7.1. Illustration of the relationship between fabric relaxation and the structural constants.

jersey fabrics, during their entire useful life, will not acquire fully relaxed dimensions.

The constants of proportionality of the fabric when taken off the knitting machine (air relaxed) and in the fully relaxed state give some guidelines to the expected dimensions after knitting and the total instability present in the fabric. These are tabulated for the five popular non-Jacquard structures in Figs. 7.2 and 7.3. Similar data can also be obtained for Jacquard designs.

It will be apparent that the constants of proportionality and consequently the fabric dimensions vary with structure, tightness, and run-in ratio. The lower the constant of proportionality for a given loop length, the longer or wider is the fabric.

Some structures such as Milano Rib and French Piqué produce wide fabrics, whilst others such as Interlock produce very narrow fabrics. Swiss Piqué and Punto-di-Roma produce 60–62 in. (152–157 cm) wide fabrics on 18 gauge 30 in. diameter machines, which is a popular

INFLUENCE OF FABRIC TIGHTNESS ON PROPORTIONALITY CONSTANTS (k-VALUES) FOR WOOL DOUBLE JERSEY FABRICS
(British units)

Structure	*Cover factor*	*k-Values*					
		After knitting		*After commercial dry finishing*		*Fully relaxed fabric*	
		k_c	k_w	k_c	k_w	k_c	k_w
Interlock	1·2	4·75	4·46	4·81	4·48		
Run-in ratio	1·28	4·83	4·40	4·87	4·42		
1:1	1·33	5·00	4·25	5·03	4·28	5·25	5·25
	1·41	5·28	3·96	5·28	4·09		
Milano Rib	1·24	4·64	3·29	4·69	3·33		
Run-in ratio	1·33	4·70	3·20	4·76	3·26		
1:1.8	1·44	4·82	3·02	4·89	3·11	5·8	3·83
	1·46	4·84	3·01	4·92	3·10		
French Piqué	1·17	5·06	3·80	5·14	3·84		
Run-in ratio	1·22	5·13	3·72	5·18	3·77		
1:2.13	1·29	5·22	3·60	5·26	3·69	6·8	4·55
	1·35	5·52	3·40	5·56	3·50		
Swiss Piqué	1·17	4·60	4·02	4·64	4·02		
Run-in ratio	1·22	4·71	3·92	4·75	3·93		
1:2.13	1·32	4·93	3·73	4·96	3·75	5·5	5·0
	1·48	5·20	3·54	5·22	3·57		
Punto-di-Roma	1·15	4·61	4·12	4·67	4·15		
Run-in ratio	1·22	4·68	4·03	4·70	4·08		
1:1	1·32	4·86	3·83	4·89	3·90	5·88	4·95
	1·48	5·28	3·53	5·30	3·63		

Fig. 7.2.

commercial width. As shown in §5.1, for the same fabric structure, width can be increased by using a finer gauge machine. Fabric width can also be increased by increasing the loop length, but this effect is generally small. In the case of a Punto-di-Roma fabric, the width can be changed significantly by altering the run-in ratio.

INFLUENCE OF RUN-IN RATIO ON PROPORTIONALITY CONSTANTS (k-VALUES) FOR WOOL DOUBLE JERSEY FABRICS
(British units)

Structure	Run-in ratio	*k-values*					
		After knitting		*After commercial dry finishing*		*Fully relaxed fabric*	
		k_c	k_w	k_c	k_w	k_c	k_w
French Pique	1·8	5·50	3·45	5·57	3·48	7·46	4·14
Cover factor	2·0	5·30	3·60	5·33	3·64	7·0	4·39
1.27	2·2	5·17	3·60	5·20	3·70	6·80	4·55
	2·4	5·15	3·60	5·20	3·70	6·80	4·55
	2·6	5·15	3·60	5·20	3·70	6·80	4·55
Swiss Pique	1·6	4·95	3·75	4·97	3·78	5·83	4·71
Cover factor	1·8	4·90	3·75	4·93	3·80	5·50	5·00
1.27	2·0	4·89	3·75	4·91	3·82	5·50	5·00
	2·2	4·88	3·75	4·90	3·82	5·50	5·00
	2·4	4·83	3·78	4·87	3·83	5·50	5·00
	2·6	4·77	3·80	4·80	3·90	5·50	5·00
Punto-di-Roma	0·8	4·75	4·02	4·80	4·07	5·98	4·80
Cover factor	1·0	4·68	4·03	4·72	4·10	5·88	4·95
1.28	1·1	4·68	4·07	4·73	4·15	5·79	4·98
	1·3	4·82	3·86	4·87	3·95	6·09	4·86
	1·5	4·85	3·75	4·91	3·85	6·24	4·63
	1·7	5·10	3·48	5·17	3·58	6·66	4·40

Fig. 7.3.

7.2 Influence of Fabric Structure on Dimensional Changes

The dimensional stability of the fabrics depend on structure, fabric tightness, and run-in ratio. Some structures such as French Piqué show high relaxation shrinkage (large changes in k_c and k_w) and some structures such as Interlock have a low relaxation shrinkage. A shrinkage of 30–35 per cent in area is common in wool double jersey structures. Increasing tightness increases width shrinkage and reduces length shrinkage without affecting the area relaxation shrinkage. Fabrics of high run-in ratio show increased instability.

7.3 Fabric Dimensions after Commercial Finishing Treatments

The majority of wool double jersey fabrics are dry finished as outlined in §6.3. The change in fabric dimensions during such finishing is small, generally 3–6 per cent in area. The change in fabric linear dimensions is generally not greater than 4 per cent. Most of this change occurs during steam relaxation and depends on fabric tension and the type of steaming equipment used.

The actual shrinkage in finishing depends to a limited extent on run-in ratio but mostly on the tightness of structure, tight structures showing higher width and lower length shrinkages than loose structures. The k_c and k_w values for various structures are shown in Figs. 7.2 and 7.3.

The dimensional changes which occur in the production of washable fabrics are only slightly greater than the changes encountered in normal dry finishing. This is because the pretreatments of pre-relaxation and setting prevent large changes occurring in fabric dimensions during subsequent processing.

Fabric weight and fabric width are two basic industrial requirements in fabric engineering and can be calculated from the k values, yarn counts, and loop length, as follows:

$$\text{Fabric weight } (\text{gm}^{-2}) = \frac{0{\cdot}0784 \times k_c \times k_w}{l} \times \text{tex count}$$

$$(\text{oz./sq. yd.}) = \frac{2 \times k_c \times k_w}{l \times \text{worsted count}}$$

$$\text{Fabric width (inches)} = n \times \frac{l}{k_w}$$

where n is the total number of needles on either the dial or cylinder of the machine.

Generally, the loop length is varied to obtain a required fabric width and the yarn count is calculated from the required fabric weight.

7.4 Other Physical Properties

Besides fabric dimensions, stability, and weight, the other major important parameter is the elastic behaviour of double jersey fabrics. Knitted fabrics, in general, show low initial stiffness and high stretch, and because of these attributes, they are more comfortable to wear than the traditional woven fabrics. However, fabrics that show high stretch invariably exhibit high residual extension or bagging. In practice, a balance has to be struck between comfort and appearance, and a good knitted fabric must exhibit a certain degree of stretch to be comfortable, but this must be limited in order to avoid excessive bagging. Structure and fibre properties influence stretch and bagging characteristics. The contribution of fibre properties varies with the conditions under which stretch and bagging are measured. Under the low loading conditions which generally occur in wear, stretch is almost entirely dependent on the structure of the fabric. This is because low stretch is accommodated by distortion of the structure, the component yarns being themselves unstrained, and fibre properties are not involved.

Increasing tightness of structure reduces stretch and, since tightness is influenced by loop length and by yarn count, it follows that reducing loop length for a given yarn, or using coarser yarn at a given loop length, will reduce stretch. For fabrics of the same tightness, stretch is inversely proportional to fabric dimensions. A fabric which is wide (low k_w) will exhibit less stretch in the width direction than a narrower fabric (high k_w). Similarly, a long fabric (low k_c) will exhibit less stretch in the length direction than a short fabric (high k_c).

8

QUALITY CONTROL

As in all textile manufacturing processes, quality control is an important feature of double jersey knitting. Since knitting is a high speed operation with a short turn round cycle, it is important that the quality control function be well organised and with clear lines of communication so that monitoring of production is precise and the need for any corrections can be signalled and acted upon quickly.

8.1 Application Points for Quality Control

Since the largest single cost in knitting is raw materials, it is essential that all yarn deliveries be sampled and tested on arrival, so that errors in count, shade, waxing, etc. can be detected and corrected before the yarn is required on the knitting machine.

Equally, quality must be checked prior to despatch to the makers-up. The normal final fabric inspection serves to check fabric faults, fabric width, weight, roll length, and probably also pattern. It may also be necessary to check on relaxation shrinkage, possibly including course length, and shade, particularly if the fabric has been piece dyed.

It is often found useful to exercise a tight control on the fabric immediately after knitting. The fabric is always inspected for faults at this stage and a simple check on fabric width and weight will sometimes detect machine setting errors as well as providing useful information for the finisher, enabling necessary adjustments to be made to the fabric properties.

It is also the responsibility of the quality control function to arrange the drawing-up of the necessary yarn and fabric specifications and in particular to provide thə knitting mechanic with the information he requires to set-up a machine. Thereafter the quality control staff should monitor the machine performance continuously, both directly on the machine, and also from inspection of the fabric produced.

8.2 Quality Control Tests

8.2.1 *Yarn Tests*

The need for, and advantages of, quality control are clearly argued in books and articles on management techniques and it is not proposed that these arguments be repeated here. Rather it will be assumed that the successful knitter will have a quality control function which is to be applied optimally to the knitting of wool yarns. Essentially, therefore, the first step is to define the points where wool yarns and fabric require special attention or differ markedly from other materials.

Like all other yarns, wool yarns are produced in batches, so that it is necessary to test on a comparative basis to detect differences which may exist between batches. In the case of wool, batches may differ in the nature of the tops used, despite the best efforts of top-makers to prevent this, and this may be significant in the final yarn. If the yarn is top dyed then there may be differences of shade to be detected.

It is often said in jest that the best test for a knitting yarn is to knit it. Like many such sayings, it contains a good deal of wisdom. A short knitting test will reveal yarn streakiness, shade variations, or excessive yarn imperfections, or winding faults far more easily than many more complex tests.

Of the normal routine laboratory tests, those which

are usually carried out on knitting yarns are yarn count, twist, strength, regularity, and moisture content. The latter is particularly important as moisture or conditioning plays an important part in the purchase price of the yarn, as discussed in §2.4.5, so that careful control of this property has important financial as well as technical consequences.

Measurement of yarn count, twist, strength, and regularity are simple tests which can be carried out on automatic or semi-automatic equipment. These properties give a good guide to the expected knitting performance and fabric appearance and can also be compared to accepted standards.

It is often argued that, for the knitter, the most important wool yarn property is the coefficient of friction and that this should be monitored. A number of instruments are available for measuring friction, and the test is relatively easy, yet its use by knitters is far from general. It would appear that knitters rely on their spinners to maintain correct yarn lubrication. Yet variation in yarn friction can quickly lead to difficult knitting conditions and poor fabric resulting from tight courses and neglect of control of this property could be serious.

Since even within a yarn batch there are sub-lots that have been processed separately in at least one stage, care must be taken that sampling is sufficient to detect differences within a yarn delivery. Stages where process differentiation could exist include top or yarn dyeing, where the machine capacities limit the size of lots dyed, and spinning, where yarn is produced on a number of independent machines.

If wool yarns have been stored for an appreciable period of time before knitting, it is recommended that their moisture content and friction be checked prior to use.

8.2.2 *Machine Tests*

After the machine has been set-up by the mechanic, the

settings and fabric should be checked by the quality control staff. Subsequent checking should be done on a regular basis to ensure that the machine conditions do not change by accident, design, or drift. Obvious items to be checked are yarn threading, run-in speeds, and tensions. Cones should also be inspected for bad winding, damage and dirt (although this may possibly be done by the knitter). A check should also be made on the machine performance, i.e. speed, the operation of stop motions, and other aids.

At the same time the fabric on the machine should be checked for obvious faults such as ladders, irregularity, incorrect shades. The take-down tension and stretcher board width will also require occasional inspection.

8.2.3 *Grey Fabric Tests*

Quality control laboratory tests on the grey fabric are limited to checks on course length, fabric width, and its weight. Normally these will only be liable to change when a machine is changed to a new pattern or re-set for some reason. The checks are run to monitor the knitting room activities and also to provide information which may be required by the finishing department or commission finishers. It will almost certainly not be necessary to check every fabric roll, and apart from machine changes, testing on a random basis will often be found to be adequate. This random basis for laboratory tests must not be confused with routine fabric inspection. This should be carried out on all fabric and faults stringed or mended as outlined in Chapter 6.

8.2.4 *Finished Fabric Tests*

Responsibility for this inspection depends on the nature of the organisation. In a vertical organisation, finished fabric tests can be integrated with and related to the earlier yarn and fabric testing. If the finishing is done on a commission basis, the situation becomes a little more complex. During finishing, the fabrics should be monitored by the finisher to ensure that the conditions required by the knitter are met. After finishing, the fabric can either be returned to the knitter or despatched directly to the customer. In the former case the knitter, who is also the vendor, can, if he so wishes, perform a final inspection before despatch. However, if the fabric goes directly from finisher to customer, there is some risk to the knitter of faulty fabric being delivered of which he is unaware.

The important features of the fabric for the maker-up are discussed in the next Chapter. These features can all be checked prior to despatch by either the knitter or the finisher. However, in addition there is information to be obtained from the final fabric which can be of importance to the knitter. The more important information is concerned with fabric dimensions, particularly width, length, and weight. Knowledge of these parameters enables the knitter to check his fabric design, and if necessary modify it as indicated in §7.3. Knowledge of fabric weight is essential to provide information on materials consumption.

In addition, other tests of a more specialised nature may be carried out on the finished fabric. These include such tests as abrasion resistance, pilling, air permeability, bagging, and stability. These are often related to fabric development rather than quality control and so are not of immediate concern in this instance.

8.3 Level of Quality Control Activity

It is very easy to become over-enthusiastic about the importance of testing in any manufacturing process. Others again would agree that testing is unproductive. However, in knitting, materials play such a large part in the total cost structure that even small savings in the use of material bring handsome returns. Circumstances vary widely between different knitters, but it has been stated that a 1 per cent change in materials losses can affect profits by up to 7 per cent. Circumstances will also affect the necessary level of quality control activity, but it does appear, at least in the British double jersey knitting industry, that expenditure on quality control —excluding inspection of fabrics for faults—averages 0·5 per cent of the expenditure on yarns. Firms producing large quantities of plain fabrics may find a lower figure than this to be appropriate, since yarn costs will form a higher than average proportion of their total costs. On the other hand, firms producing patterned or speciality fabrics may require a higher level, their yarn costs forming a lower proportion of their total costs due to the greater importance of processing and development costs.

9

MAKING UP

Clothing manufacturers, some of whom have previously been involved only with woven fabrics, now must work with fabrics produced in double jersey structures and, as a result, have experienced some problems. Some of these are associated with properties of knitted fabric, such as thickness, stretch, and running or broken threads, while others result from the properties of the fibre used, e.g. needle heating and melting of the fabric.

9.1 Preliminary Inspection, Testing and Storage

It is recommended that all fabric should be inspected by the maker-up on delivery. Any faults should be marked according to the possibility of mending, and any variations in fabric appearance should be noted, i.e. wrong pattern, skewing, width variations, and shading.

Steam press shrinkage should be measured. This is carried out by marking out an area approximately 25 cm square on a sample and steaming on a steam press, with the head raised, for 30 seconds, followed by vacuum.

Any discrepanies in the fabric should be taken up immediately with the knitter to prevent continuation of the fault, and to have the observed fault corrected or the faulty fabric replaced.

After mending, fabrics should be stored in open width (to prevent centre marking) in a store which is at fairly constant humidity. The rolls should be stored horizontally, as vertical storage can result in slippage and distortion of the fabric. Rolls of fabric should not be stored one on top of another since pressure between the rolls can cause considerable distortion to occur.

9.2 Garment Design

A feature of double jersey fabrics is that they exhibit higher stretch in both directions than woven fabrics when placed under a given load. Since the degree of comfort of a garment largely depends on the pressure which it exerts on certain critical body parts, it follows that knitted fabrics are more comfortable from this point of view than woven fabrics. It also implies that, within reason, garments can be made more closely fitting than previously, and still be comfortable to wear.

In certain garment areas, excessive stretch is a disadvantage. For example, in trouser waistbands, double jersey fabrics should be reinforced with a rigid woven fabric.

9.3 Laying-up and Cutting

As double jersey is easily damaged and distorted, it is important that handling should be carried out with care. All projections which could possibly cause snagging should be removed or smoothed down.

Laying-up should be carried out under minimum tension and garment patterns must be placed on the fabric so that hems point to one end of the lay only. This must be followed by cutting in one direction, starting from the hem each time and cutting down both sides of each part. This procedure minimizes garment size variation.

The number of plies in a lay is not restricted except by commercial considerations.

Most cutting systems require the preparation of a marker trace. This is prepared from the master card patterns by, for example, tracing the outline of the pattern onto paper. This paper, or marker trace, is fastened securely on to the top layer of fabric in the lay with pins or staples. It is important that pins and staples should be inserted only in waste areas of cloth in order to prevent damage.

When chalk is used, the normal dry type is preferred to wax chalks, which leave stains. Drilling to locate pockets should be checked to ensure that this does not cause damage to the fabric.

It is recommended that the lay should be allowed to relax prior to cutting. Band and straight knives should be kept in good condition and are recommended in preference to circular knives, except for separating or rough cutting. Notching should always be done with a straight knife or notcher. Pocket mouths should not be cut with a knife before sewing starts.

The ticketing of parts is best done with the heat-seal type of label, thus preventing the possibility of damage by sewing or stapling. The labels must be attached on the reverse side of the fabric.

9.4 Seaming

Seams in double jersey should be more extensible than seams in woven fabrics. If seam extensibility is inadequate, "cracking" may occur, i.e. the two parts of the seam may become separated because the sewing thread has broken and the stitch has been unable to lock the broken ends.

The seam type normally used is SSa-1, sewn on a single needle lock stitch machine (stitch type 301). A double "locked" chain stitch, type 401, may be used in certain cases but it suffers from the increased likelihood of cracking of the seam if a thread breaks. The increased stretch that is so often claimed for this type of stitch is of little significance, especially if the thread recommendations for the lock stitch are followed as set out below.

Fraying out of seams is not usually a problem but manufacturers should experiment to determine if over-locking is necessary.

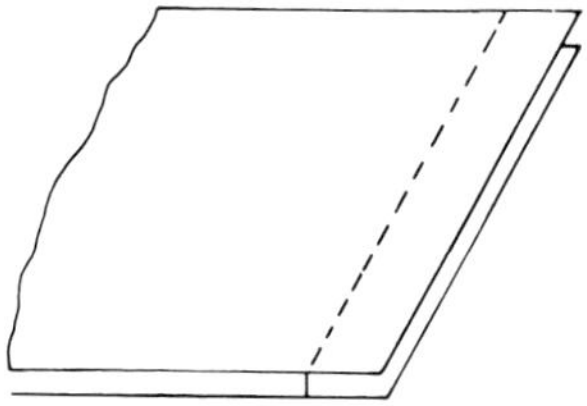

SEAM SSa-1

STITCH 301

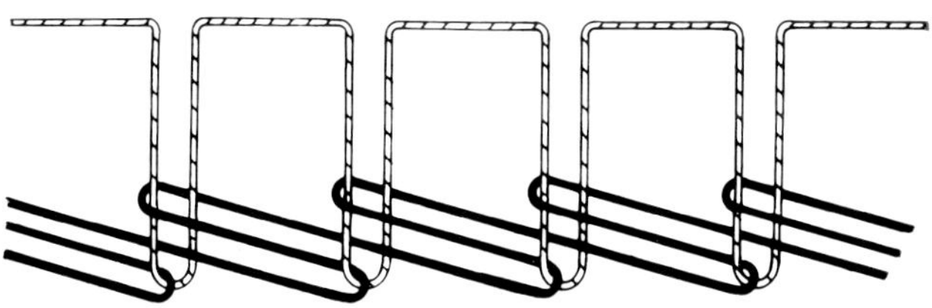

STITCH 401

Fig. 9.1. Seam and stitch types recommended for double jersey fabrics.

9.5 Sewing Damage

Experiments have shown that provided reasonable precautions are taken, sewing damage is normally no problem in all-wool fabrics. Generally the finer the yarn count and the finer the gauge, the greater the possibility of damage.

Dry-cleaning and needle size are, however, the most important factors affecting sewing damage, and high sewing speeds can also cause an increased fault rate.

When fabrics are dry-cleaned to remove oil and dirt after knitting, or when solvent shrink-resist treatments are applied, they show an increased tendency to damage, particularly in the cases of finer counts and tightly constructed fabrics. It is therefore desirable to apply a suitable lubricant after dry-cleaning.

To minimize the possibility of damage, it is recommended that the amount of machining or needlework be kept to a minimum. Hence, for example, fusible interlinings should be used instead of conventional interlinings which require to be pad stitched.

Finer sewing needles produce less damage than coarser needles and fabrics should be sewn with a medium ball-point needle, size 12 (Singer) (metric size 80), using approximately 12 stitches per in. (4·5 stitches per cm).

It is also most important that needles are inspected for wear and are discarded when unsuitable, though it may be more economical to change the needles at regular intervals in any case.

9.6 Seam Pucker

Seam pucker is rarely a problem in wool double jersey fabrics, but it is advisable to ensure the correct adjustments of thread tensions and presser foot pressure.

Sewing machines are adjusted with two plies of fabric under the presser foot, with the feed dogs below the throat plate. Foot pressure is increased or decreased until it is possible to move the fabrics backwards and forwards smoothly by hand. Excessive presser foot pressure will cause spreading of double jersey fabric under the area of the presser foot. If fabrics are sewn under these conditions, when the seamed fabrics leave the presser foot there will be some recovery towards the original dimensions, causing puckering.

Having adjusted the foot pressure, it is necessary to adjust the feeding mechanism, and this is usually at an optimum when the feed dogs are adjusted so that, at their maximum height, $\frac{1}{2}$–$\frac{3}{4}$ of the teeth are showing above the feed plate. The most suitable type of feed dogs are those with two long rows of teeth, one on either side of the needle, and two short rows, one before and one after the needle. Machinists should not pull or push work during sewing or the needles will be damaged more often.

Correct thread tension in a lock stitch is achieved by adjusting the bobbin tension so that the bobbin and bobbin case will just fall when suspended under their own weight by the thread. This is usually 15 g tension when measured by a yarn tension meter, with the bobbin and case in the machine and the thread running through the hole in the throat plate. The top or needle thread tension is then adjusted with fabric in the machine to give a balanced seam (about 150 g tension). A test run on the fabric to be sewn should always be made and, if necesssry, a precise measure of the lengths of thread in both parts of the seam should be compared.

9.7 Sewing Thread

The choice of the most suitable sewing threads for joining two fabrics together depends on the nature of the fabrics concerned. The greater the fabric is extended in use, the greater the stress on the sewing thread, which is always less extensible than the fabric. In most stitch types there is one point in the sewing where two different yarns cross, and it is at this point where the force generated by the given extension is concentrated. Consequently the stress in the thread is higher for finer threads, and is frequently sufficient to cause rupture. An easily extensible, fairly coarse thread is the best sewing thread for double jersey fabrics for any given stitch density.

9.8 Shape Stabilization

Woven fusible interlinings are recommended wherever it is necessary to prevent excessive stretch and distortion in a knitted garment during wear, e.g. jacket forepart, fly area, pocket flaps, and welts. The interlining should be chosen so as to be compatible in weight with the outer fabric and should ideally produce a laminate of the minimum weight required to obtain the desired performance. The bond strength should be the maximum obtainable, compatible with hand requirements and strike through of adhesive onto the front of the garment.

Waistbands are a particularly important case where stabilization must be carried out by fusing to a suitable woven interlining.

9.9 Pressing

Care must be taken not to impose excessive stretch when placing the garment on the press. Soft cladding must be used to minimize embossing, which can be a problem in certain qualities of fabric.

When pressing jackets or coats, several presses will be used. In each press, it is preferable to use light pressure with minimum steam, followed by at least 10 seconds vacuum. Press covers should be changed frequently to avoid staining, and all the work surfaces must be free from sharp edges.

9.10 Creasing

Permanent trouser creases may be inserted, if required, into all-wool double jersey garments in one of two ways—by chemical setting or by sewing-in.

9.10.1 *Chemically Set Creases*

The crease lines are sprayed with a solution containing:

MEAS (Monoethanolamine sesquisulphite)
50 g (40 ml) per litre
Nonionic wetting agent
1–2 g (1–2 ml) per litre

Stainless steel, plastic, or glass utensils only should be used for making-up such solutions. The fabrics should have previously been tested for colour fastness to this solution.

It is important to obtain the correct pick-up, which should never be less than 40 per cent. Trials should be carried out to determine the correct technique to obtain 40 per cent pick-up, and this is easily done if a special spray booth is available. Hand spraying may be carried out, but great care is necessary to ensure correct and even application.

The garments should not be allowed to dry out before pressing as adequate setting is obtained only if water is present at the time of creasing.

The pressing sequence is as follows:

20 sec. steam	head closed
20 sec. bake	head closed (no steam)
10 sec. vacuum	head open
Normal steam pressure	60 lb. in.$^{-2}$ (4 kg cm^{-2})
pressure	2 atmospheres

The minimum pressure necessary to obtain the desired sharpness of crease should be used as excess pressure or too localized spraying will result in glazing on the crease. The creases inserted by the above procedure are very difficult to remove after pressing, so that extreme care should be taken in their correct location.

The garments should then be dried by hanging vertically or laying flat.

9.10.2 *Sewn-in Creases*

Creases can be sewn in by first locating the crease line with a short steam press, followed by sewing down the crease line from waistband to cuff using a lockstitch machine.

The sewing thread for creases must be of high extensibility so that it does not break when the crease is stretched during wear.

10

SOME TYPICAL WOOL DOUBLE JERSEY FABRICS

10.1 Plain Fabrics

The three most popular plain structures are Interlock, Swiss Piquè, and Punto-di-Roma.

10.1.1 *Interlock*

Although some ladies' dress and blouse fabrics are made in Interlock construction, the majority of wool Interlock fabric produced is for underwear.

The fabric is produced on circular machines of 10–30 in. diameter to give the required body size. The popular machine gauge is 18, although 14 and 24 are also used. Depending on the gauge of the machine, the yarns used vary from 18 to 36 tex.

The fabric is generally knitted to a cover factor of 1.3. It is almost always wet finished on a winch. The procedure entails scouring, bleaching, and perhaps oxidative shrink-resist treatment.

10.1.2 *Swiss Piqué*

This is one of the popular ladies' dress fabrics, knitted mainly on 18 gauge circular double jersey machines using 28 tex or coarser yarn.

The fabric is knitted to a cover factor of 1.3, which is found to give 38 courses per in. when measured on one side of the fabric at a run-in ratio of 1:2.2. The fabric weight is generally 280–310 gm^{-2} and if knitted on a 30-in. diameter machine the fabric is dry finished to 60 in. (152 cm) width.

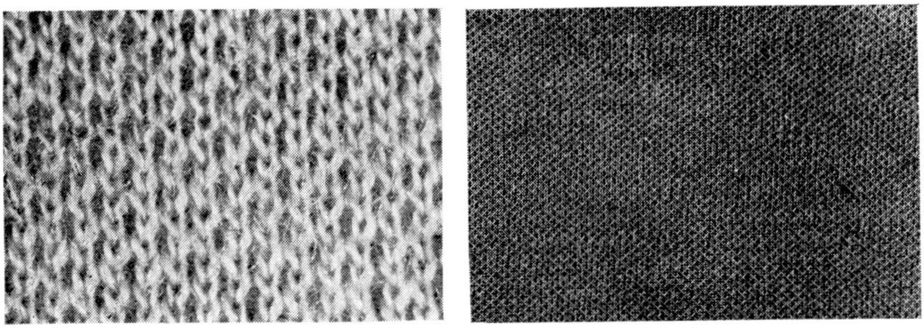

A

B

Fig. 10.1. Popular plain double jersey fabrics. (A) Swiss double pique (B) Punto-di-Roma.

10.1.3 *Punto-di-Roma*

This is perhaps the most popular of non-Jacquard structures and is used for both ladies' wear and men's wear. The pronounced rib appearance is common for ladies' wear, whereas flat appearance is common in men's wear. These appearances can be obtained by varying run-in ratios, higher run-in ratios giving a more pronounced rib.

Fig. 10.2. A flat Jacquard fabric (above) and a blister Jacquard fabric (below).

Common construction details are as follows:

	Ladies' Wear	*Men's Wear*
Yarn count, tex (worsted count)	28 (32)	28 (32)
Fabric cover factor	1·28–1·35	1·4–1·5
Run-in ratio	1:1·7	1:1·3
Courses per inch	38	45
Fabric width, inches* (cm)	66 (168)	60 (152)
Fabric weight, gm^{-2}	300	300–350

* These figures are for a 30 in. diameter, 18 gauge machine.

A 22 gauge machine would give a greater width in the ratio of 22:18.

10.2 Jacquard Fabrics

Jacquard fabrics are generally difficult to categorize in terms of cover factor or loop length. Generally, they are made from 28 tex (32's worsted count) yarn to a weight of between 280–300 gm^{-2}

The courses per inch of Jacquard fabrics are normally measured on the back of the fabrics. The typical values are:

	Fabric Face	*Fabric Back*
Two-colour Jacquard	38	38
Three-colour Jacquard	28	42
Four-colour Jacquard	24	48

Contrary to general belief, the weight and the width differences between two-, three-, and four-colour Jacquards are very small.

10.3 Blister Fabrics

To produce a blister, the loops are held on the dial needles during the knitting of additional cylinder courses and consequently the knitting performance is considerably inferior to Jacquard fabrics. Only a limited amount of blister fabric is commercially produced, two-colour blisters being the most popular.

A typical blister fabric is made from 28 tex yarn to a weight of 280–340 gm^{-2}. There are 35 courses per inch, slightly less than for a corresponding two-colour plain Jacquard fabric.

APPENDIX I

COUNT CONVERSION TABLE

Tex	*Nm*	*Cotton*	*Worsted*	*Denier*
15·0	66·7	39·4	59·1	135·0
17·5	57·1	33·7	50·6	157·5
20·0	50·0	29·5	44·3	180·0
22·5	44·4	26·2	39·4	202·5
25·0	40·0	23·6	35·4	225·0
27·5	36·4	21·5	32·2	247·5
30·0	33·3	19·7	29·5	270·0
35·0	28·6	16·9	25·3	315·0
40·0	25·0	14·8	22·2	360·0
45·0	22·2	13·1	19·7	405·0
50·0	20·0	11·8	17·7	450·0
55·0	18·2	10·7	16·1	495·0
60·0	16·7	9·9	14·8	540·0

APPENDIX 2

TRANSLATION OF COVER FACTOR AND RUN-IN RATIO INTO COURSE LENGTHS

Before a machine can be set-up to produce a plain double jersey fabric, it is necessary to translate the structure parameters of cover factor and run-in ratio into course lengths.

Starting from a loop diagram of a given structure, the number of loops in each course of the repeat unit are counted.

If p = the number of loops formed in the dial *or* cylinder (plain) courses in the width of the repeat unit,

q = the number of loops formed in the dial *and* cylinder (rib) courses in the width of the repeat unit,

l_1 = the length of yarn in one loop of the dial *or* cylinder (plain) course,

l_2 = the length of yarn in one loop of the dial *and* cylinder (rib) course,

Then as defined in §3.1.2 and §3.1.3.

Run-in ratio, $1{:}r = pl_1{:}ql_2$ (3a)

and average loop length, $l = \dfrac{pl_1 + ql_2}{p+q}$ (3b)

From equations (3a) and (3b), the individual loop lengths can be expressed as:

$$l_1 = \frac{(p+q)\,l}{(l+r)\,p} \tag{3c}$$

and $$l_2 = \frac{(p+q)\,l \times r}{(l+r)\,q} \tag{3d}$$

The length of yarn in one dial or cylinder (plain) course,

$$X = l_1 \times p \times \frac{n}{m} = \frac{(p+q)}{(l+r)} \times l \times \frac{n}{m} \tag{3e}$$

and the length of yarn in one dial *and* cylinder (rib) course,

$$Y = l_2 \times q \times \frac{n}{m} = \frac{(p+q)}{(l+r)} \times l \times r \times \frac{n}{m} \tag{3f}$$

where m is the number of needles in the width of the repeat and n is the total number of needles on the dial or cylinder (i.e. half the total number of needles on the machine). n/m is thus the number of repeat units in the fabric width.

If the requirements are given in terms of cover factor rather than loop length, then equations (3e) and (3f) become

$$X = \frac{(p+q)}{(l+r)} \times \frac{n}{m} \times \frac{1}{\text{cover factor}} \times \frac{1}{\sqrt{N}} \tag{3g}$$

$$Y = \frac{(p+q) \times r}{(l+r)} \times \frac{n}{m} \times \frac{1}{\text{cover factor}} \times \frac{1}{\sqrt{N}} \tag{3h}$$

where N is the yarn count (worsted).

Similarly on the tex system,

$$X = \frac{p+q}{l+r} \times \frac{n}{m} \times \frac{\sqrt{T}}{\text{tightness factor}} \tag{3j}$$

$$Y = \frac{p+q}{l+r} \times r \times \frac{n}{m} \times \frac{\sqrt{T}}{\text{tightness factor}} \tag{3k}$$

where T is the yarn count in tex.

For practical knitters, these equations are reduced to simple forms for specific structures, as shown in Fig. A.2.1.

STRUCTURE		COURSE LENGTH, inch	
NAME	DESIGN	CYLINDER OR DIAL	CYLINDER AND DIAL
INTERLOCK	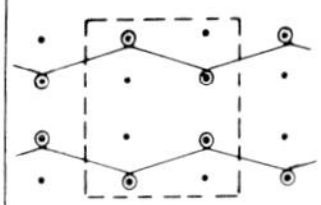		$\frac{n}{2 \times (cf) \times \sqrt{N}}$ $\frac{n \times \bar{l}}{2}$
MILANO RIB	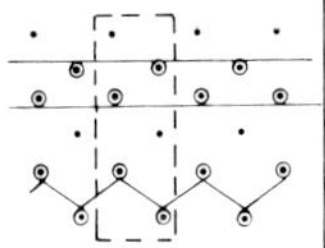	$\frac{3n}{2(1+r) \times (cf) \times \sqrt{N}}$ or $\frac{3n\bar{l}}{2(1+r)}$	$\frac{3nr}{2(1+r) \times (cf) \times \sqrt{N}}$ or $\frac{3n\bar{l}r}{2(1+r)}$
PUNTO-DI-ROMA	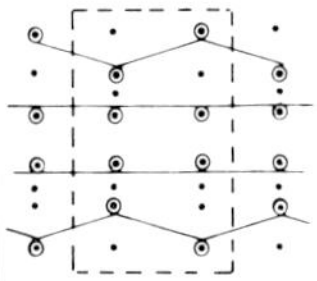	$\frac{n}{(1+r) \times (cf) \times \sqrt{N}}$ or $\frac{n\bar{l}}{(1+r)}$	$\frac{nr}{(1+r) \times (cf) \times \sqrt{N}}$ or $\frac{n\bar{l}r}{(1+r)}$
SWISS PIQUE	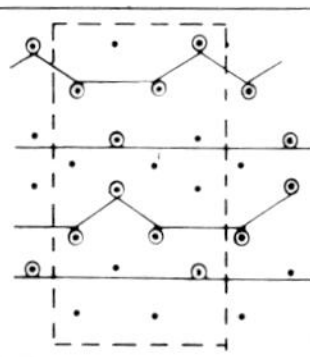	$\frac{n}{(1+r) \times (cf) \times \sqrt{N}}$ or $\frac{n\bar{l}}{(1+r)}$	$\frac{nr}{(1+r) \times (cf) \times \sqrt{N}}$ or $\frac{n\bar{l}r}{(1+r)}$
FRENCH PIQUE	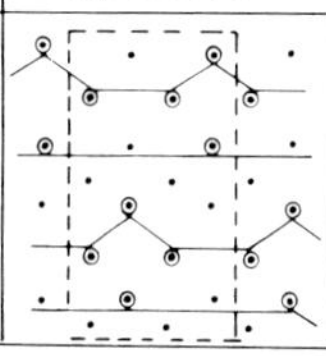	$\frac{n}{(1+r) \times (cf) \times \sqrt{N}}$ or $\frac{n\bar{l}}{(1+r)}$	$\frac{nr}{(1+r) \times (cf) \times \sqrt{N}}$ or $\frac{n\bar{l}r}{(1+r)}$

N - Number of needles in the machine (cylinder or dial)

N - Yarn count, worsted

$\bar{l}$ - Average loop length, inches

(cf) - Fabric cover factor

1:r - Run in ratio

Fig. 4.2.1. Calculation of course lengths for some plain double jersey structures.

INDEX